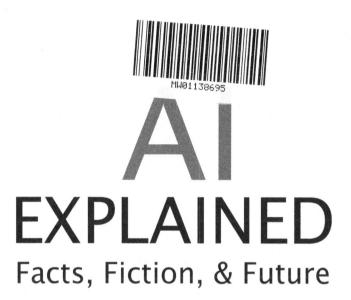

AI
EXPLAINED
Facts, Fiction, & Future

Nigel Poulton

© 2024 Nigel Poulton Ltd.

To my mum.
For being a brave single mother and never giving up.

To my dad.
*For buying me my first computer and encouraging
me to pursue a career in technology.*

Contents

1. Primer 1
2. Building advanced AIs 27
3. Using AI 45
Part 2 65
4. Humans misusing AIs 67
5. Rogue AIs 79
6. Conclusion 99
Glossary 103
Endnotes 109
Index 115

Acknowledgments

The old proverb that *"it takes a village to raise a child"* is equally true of writing books.

Fortunately, this book was raised and nurtured (written and edited) in a fantastic *village* and I'm grateful to every single person who's helped.

I wish I could thank everyone by name as I've had fantastic help from family and friends who've reviewed chapters, checked for spelling mistakes, told me when parts were boring or didn't make sense, given their opinions on fonts, designed the cover, and even helped choose the title of the book. I couldn't have done it without you, but it was a lot of fun and I'm extremely proud of the book we've created.

I must thank my wife and children for their patience and support. Not only do they allow me the time and space to write, but they also keep me real and bring an important balance into my life. As a quick example, I'm terrible at managing my time, and if it weren't for them, I'd probably spend all day writing my books. So, taking time out to watch the *Formula 1* and *The Rookie* with Lily, coach Abi's football, take Mia to her swimming events, and have lunch dates with my wife Jen has brought a balance to my life that has helped me and the book.

Thank you!

About the author

Nigel Poulton is a best-selling author and video-trainer. He has sold over 1.5 million books and video training courses helping software engineers and computer scientists learn to use new technologies. He is best known for his ability to demystify and explain complex topics in simple terms.

Before writing books, Nigel deployed and managed cutting-edge internet technologies in some of the most demanding business environments in the world.

He lives in the UK with his wife and three children, where he coaches youth sports and enjoys watching football (soccer) and Formula 1.

You can contact Nigel via the contact form on his website.

@nigelpoulton
nigelpoulton.com

Preface

They don't know it, but I've written this book for my mother-in-law Val and my Uncle Tony. Well... for them and the millions of other fantastic human beings curious about artificial intelligence (AI).

Val is a well-educated former school teacher, mother and grandmother to more than I can count, and an active member of her local community. Tony is a retired Royal Air Force (RAF) pilot who enjoys caravanning around the UK and France with my Auntie Ann. He was awarded an MBE (highest award in the Order of the British Empire) for his many years of service to his local communities.

I've organized the book so you can start on page one and read to the end in a single day. It's also fun and engaging, meaning once you start reading, you won't want to stop.

You'll learn the fundamentals of AI, some of the exciting ways we're using them today, and the kinds of things to expect and look out for in the future. And the book doesn't shy away from the hard and sometimes controversial questions.

I've also tried hard not to impose my own opinions. I present the facts, pose some fascinating questions, and leave you to form your own opinions. Whether you want to use AI, avoid it, be excited about it, or be afraid of it is entirely up to you.

One last thing. Except for spelling and grammar checks, I haven't used AI to help with any of the writing aspects of this book–it's all written by a *good old human being.*

1. Primer

This chapter will lay some important groundwork and get you up to speed with the basics of artificial intelligence (AI). It will also pose some questions to stimulate your thinking. Don't worry if some things are confusing or you feel like you want to know more; we'll revisit everything in more detail in later chapters.

We'll begin by discovering what AI means to you and then start building on that. We'll introduce you to some of the world's most advanced AIs, give you a glimpse of how we create them, and round out the chapter by asking some important questions and outlining a few concerns.

At the end of each chapter section, I'll list an important *take-home point* and build this out as we progress through the chapter.

Remember, though, the goal of this chapter is to get you started.

In the beginning

Little did they know it, but more than two and a half million years ago, when our early ancestors created the first stone tools, they set us on a voyage of change and discovery where every subsequent tool and invention has edged us ever closer to the ultimate tool... AI!

While that might sound grandiose, a quick look at a few pivotal tools and inventions reveals a clear pathway to AI.

From stone tools, we discovered metalwork, which gave us better tools for hunting and agriculture. Fast-forward a few millennia, and we have the scientific method, which gifted us, among other things, germ theory, astronomy, and modern medicine. Fast-forward again to the 20th century, where the invention of the transistor enabled modern computers, which, in turn, enabled us to invent the internet, mobile phones, and now AI.

Just like the wheel, modern medicine, and the internet, AI has the potential to change the course of human civilization and is here to stay.

But are we on a collision course with AI, and what will happen when we build advanced AIs?

Undoubtedly, AI *could* be our most disruptive invention to date–more disruptive than the internet, mobile phones, and social media. As such, it could ignite a golden era of human progress and prosperity or a world of hardship and suffering. But even bigger questions exist, such as whether AI could advance so far beyond humanity as to relegate us to a mere footnote in the broader evolution of intelligence.

Of course, there's also the possibility that AI will fail to live up to the hype and be nothing more than a footnote in human history.

These are big questions, and there are many more like them. And while we can't predict exact futures, this book will give you the knowledge and confidence to form your own informed opinions.

Take home point: We're building AIs.

What is AI

Before I throw the dictionary definition at you, ask yourself the following questions. It might be interesting to write down your answers to see if they change by the end of the book.

- What's the first thing that comes to mind when you hear the term "AI"?
- How would you describe AI in one sentence?

Now, compare your answers with some of the answers I got when I asked my family and friends the same questions.

What's the first thing that comes to mind when you hear the term "AI"?

- Robots (several people gave this answer)
- ChatGPT
- Machines/computers processing large amounts of information, but not without mistakes
- Assistance to do things
- Computers making decisions
- Something that steals identities
- A computer system that learns from itself

Describe AI in one sentence.

- Computers that can learn by themselves
- A system that will learn and eventually give better output
- Software that learns and develops based on information given to it
- A capability to make decisions like a human would, but a lot more powerful
- Machines to replace people

While none of the people I asked are AI experts, their responses are interesting and demonstrate varying attitudes and levels of understanding. For example, some didn't feel they could even attempt to describe AI.

Now, let's ask a few so-called experts to describe AI in a single sentence.

I asked the popular Merriam-Webster online dictionary, I asked Google, I asked two of the world's most advanced AIs, and I asked myself. Here's what I got:

Merriam-Webster online dictionary: *"The power of a machine to imitate intelligent human behavior."*

Google's top answer: *"AI is technology that enables computers and machines to simulate human intelligence and problem-solving capabilities."*

Claude AI: *"AI is the simulation of human intelligence in machines, enabling them to learn, reason, and perform tasks that typically require human cognitive abilities."*

ChatGPT: (Unfortunately, ChatGPT replied with one of the longest and most complicated sentences I've ever seen. In fact, it was so long and technical that I didn't have the patience to read it).

Me: *"Machines with human-like intelligence or better."*

There are some interesting trends in these "expert" responses.

Every answer referenced *human intelligence*, and mine was the only one that didn't include the terms *imitate* or *simulate*. Mine was also the only answer to suggest AIs might have greater intelligence than humans. None of the answers implied any form of consciousness or self-awareness.

Take home point: We're building AIs, which are machines with human-like intelligence.

From spam filters to superintelligence

Not all AIs are created equal.

To help classify them, researchers group AIs into one of the following three classes[1] based on their level of intelligence:

- Artificial Narrow Intelligence (ANI)
- Artificial General Intelligence (AGI)
- Artificial Superintelligence (ASI)

To keep the jargon and acronyms to a minimum, we'll call them *narrow intelligence, general intelligence,* and *superintelligence.* And yes, writing *superintelligence* as a single word is normal.

In the simplest terms, narrow intelligences are the least intelligent AIs, superintelligences are the most intelligent, and general intelligences are somewhere in the middle.

At the time I'm writing this book, we've only created narrow intelligences. However, things are moving fast, and some researchers think we're close to creating general intelligences, and once we do that, superintelligences will quickly follow.

Let's take a closer look at each.

All of the AIs that we have today are narrow intelligences. They're the most basic kind, and we sometimes refer to them as *weak intelligence.* They include things like Alexa, Siri, your email spam filter, facial recognition systems, chess engines, self-driving cars, fraud detection systems, photo editing apps on your phone, and even ChatGPT. Many narrow intelligences can only do a single task, and none of them can learn beyond their original programming. This means they cannot learn from their experiences and get smarter.

General intelligences are the next level up and are a form of *strong intelligence* that can understand and learn like humans. We haven't invented them yet, but nations, investors, and global technology companies are investing hundreds of billions of dollars in a race to be first. If and when we create them, they will be able to do everything an educated adult human can do, **including learning beyond their original programming**. For example, a general intelligence programmed to specialize in mathematics might teach itself biology and medicine and help us invent cures for diseases. General intelligences may even invent things themselves without collaborating with humans.

Superintelligences will be the real game changers, as **they'll operate far beyond human intelligence** and have almost unlimited potential. For example, they may one day eradicate disease, invent clean power, reverse climate change, and solve every solvable problem. They'll also have immense potential to cause harm and suffering.

Now, I know how ridiculous some of those superintelligence predictions may sound. They used to sound just as unrealistic to me. However, if we accept even the remotest possibility that artificial superintelligence may one day be orders of magnitude smarter than the smartest human, we must also accept that today's rules will no longer apply. We'll talk about this a lot in later chapters.

However, we're getting ahead of ourselves as AGIs and ASIs are purely hypothetical at the time of writing. But as previously stated, some experts[2] believe we're on the verge of creating general intelligences–possibly as soon as 2030 or before. And suppose they're right and we successfully create general intelligences that can learn and improve. In that case, we may enter a cycle where AIs improve themselves at ever-increasing rates until we have an *intelligence explosion*[3] that gives birth to a superintelligence. If this happens, and it's a big *"if"*, but if it happens, all bets are off, and we'll be in uncharted territory. In the words of the late Vernor Vinge,[4] *"We will soon create intelligences greater than our own. When this happens... the world will pass far beyond our understanding."*

To add some balance to the discussion, many other experts[5] think we're nowhere near creating general intelligences, and some think[6] we may never create them.

Take home point: We're building AIs, which are machines with human-like intelligence that *may* one day outsmart us.

What is ChatGPT

As I've mentioned ChatGPT a few times, I guess I should explain it.

ChatGPT is an AI created by a company called OpenAI and is responsible for sparking much of the current public interest in artificial intelligence. People are using it to write letters and essays, summarize large documents, answer questions, and my mother-in-law is even using it for recipe ideas and more.

It's a type of AI called a *chatbot* and is currently a narrow intelligence. I say it's "*currently* a narrow intelligence" because it may one day evolve into a general intelligence and even a superintelligence. In fact, some people already class it as a *potential emerging AGI,* which is jargon for saying that ChatGPT is constantly being improved and may one day be far more advanced than it is today.

But wait… if ChatGPT is a narrow intelligence, and we just said narrow intelligences can't learn, how can it be constantly improving?

That's an important question we'll return to later in the book. But here's the quick answer. Even though narrow intelligences cannot learn and improve themselves, the companies that created them are constantly making newer, more intelligent versions that can give the impression that ChatGPT and other AIs are getting smarter. As a quick example, I've used ChatGPT for a while, and it's definitely improved over time. However, this isn't because the original version of ChatGPT has learned new things for itself. It's because the company that owns ChatGPT has created and released improved versions. If today's version of ChatGPT could improve itself without OpenAI making new versions, it would be a *general intelligence*.

I also said ChatGPT is a *chatbot,* so what is one of those?

Chatbot is short for *chat robot* and describes a type of AI we can have human-like conversations with. But don't read too much into the *robot* part; ChatGPT isn't your classical metal robot that looks like a human. It's a sophisticated computer program running on thousands of computers all over the planet.

There are lots of ways you can talk to ChatGPT, including the chatgpt.com website, and apps for your phone and computer. However, many everyday things already use ChatGPT behind the scenes without us even knowing. For example, Microsoft has invested more than $10B in ChatGPT and uses it to power most of the AI features in Windows 11. Apple also uses ChatGPT behind the scenes to power some of its own tools, including Siri, its Messages app, email, phone call transcripts, and many of its writing tools.

No matter how they implement it, there's a clear trend towards everyday tools leveraging advanced AIs behind the scenes. As such, we shouldn't be surprised if more and more devices like Alexa, smart TVs, wearable devices like sunglasses, phone apps, websites, helpdesks, and search engines all use AI behind the scenes.

All of this suggests you may have already had conversations and interactions with chatbots like ChatGPT without even knowing. And everything is trending toward this becoming the norm, so don't be surprised if AI becomes as integral to everyday life as the internet.

Here's a transcript of a quick conversation I had with ChatGPT using the ChatGPT app on my computer. If it feels like I'm asking silly questions, it's because I am. ChatGPT is capable of far more complex and in-depth conversations; I just thought this one would be a good introduction.

Me: *Can you explain the offside rule to me?*

ChatGPT: *In soccer, a player is offside if they are nearer to the opponent's goal line than both the ball and the second-last opponent (usually the last outfield player) at the moment the ball is played to them, unless they are in their own half of the field.*

Me: *Why did you give me the soccer definition rather than another sport?*

ChatGPT: *I provided the soccer definition of "offside" first because it's commonly associated with soccer and frequently asked about due to its complexity and impact on the game. If you're interested in how "offside" applies in another sport, I can explain that too!*

Me: *Why did you call it soccer instead of football?*

ChatGPT: *I used "soccer" to ensure clarity for all readers, as "football" refers to different sports depending on the country.*

Me: *OK. What is a pole?*

ChatGPT: *A pole is a long, slender cylindrical object typically made of wood, metal, or plastic.*

Me: *What is a Pole?*

ChatGPT: *A Pole refers to a person from Poland, an ethnic group native to the country of Poland in Central Europe.*

If you look closely enough, you'll find a lot of nuance in the conversation. Here are just a few examples.

ChatGPT assumed I was referring to the soccer *offside* rule and used sound logic to call it soccer rather than football (much to my chagrin as a British person who calls it "football"). It also rightly or wrongly guessed that I was asking about the physical object when I wrote *pole* with a lowercase "p", but guessed I meant the nationality when I used an uppercase "P". It also gave decent answers to the questions. However, if I'd been talking to a human, I'd have expected a more conversational response to the repeated question about the Pole, such as *Oh, sorry, did you mean a Polish person in your previous question* or *Are you asking about the nationality now?* Without this kind of nuance, AI conversations can sometimes feel a little robotic.

There are also many other advanced chatbots, such as Claude from Anthropic and Gemini from Google. All of these are a class of AI called *generative AI (GenAI)* which tells us they're capable of *generating* new, unique content. We've already seen ChatGPT generate human-like speech, but most GenAI chatbots can also create images, videos, music, poetry, and more.

As a quick example, I asked ChatGPT and Claude to each *write a four-line poem about Neil Armstrong* and got the following.

ChatGPT:

> *Neil Armstrong took a giant leap,*
> *Stars and moons his to keep,*
> *With one small step on lunar sand,*
> *He bridged the skies from land to land.*

Claude:

> *One small step for a man, they say,*
> *Armstrong's boot touched lunar clay.*
> *A giant leap for humankind,*
> *His moon walk etched in history's mind.*

Both took less than two seconds to generate, and both correctly assumed Neil Armstrong the astronaut and not my high school sports teacher with the same name.

I'll leave you to decide which is the better poet. But before you dismiss them both as utterly hopeless, try creating something better yourself in under a minute. I tried and failed.

If you can't do better, what does that tell you about the current abilities of AI? And remember, there are specialized AIs that create music, art, and videos that may blow your mind. And they're constantly improving.

At the time of writing, ChatGPT, Claude, and Gemini represent the pinnacle of AI research but are still narrow intelligences. But, as previously stated, some researchers consider them *emerging general intelligences.*

In summary, ChatGPT is an AI capable of human-like conversations. It can also create new, unique content such as prose and verse, computer programs, and images. Future versions will be able to make music, videos, and more.

Take home point: We're building AIs, which are machines with human-like intelligence that *may* one day outsmart us. However, today's state-of-the-art chatbots like ChatGPT are just the beginning.

How do we create AIs like ChatGPT

AI research is one of the most rapidly advancing scientific fields and shows no signs of slowing down. This means that what we consider state-of-the-art today will feel antiquated tomorrow. As such, we won't go into detail here as the methods we use to train AIs are changing fast. However, at a high level, creating AIs like ChatGPT involves three main steps:

- Create the AI
- Train the AI
- Release the AI

Almost all of the learning happens during the training stage, and once released to the public, an AI's intelligence is fixed and does not improve. Yes, you can feed some chatbots more knowledge and give them more experiences, but their intelligence that processes the data is fixed. Currently, the only way for AIs to increase their intelligence is to train and release a new version. This will change if we create *general intelligences* as these will learn from their experiences and continuously improve themselves.

Training AIs like ChatGPT involves four main steps:

1. Choosing a dataset
2. Pre-training
3. Fine-tuning
4. Evaluation

Choosing a dataset. The first thing an AI needs is a massive dataset from which to learn. This means giving it as many books, articles, websites, and other language-related items as possible. The more you give it, the more it learns and becomes better at understanding human language.

Pre-training. The AI then trains itself by analyzing the dataset and learning the basic structure of human language, such as words and sentences. It also learns how to create its own content by predicting the missing words in the middle or at the end of sentences.

Fine-tuning. Once trained, you fine-tune the AI by training it on more specific datasets and grading its responses. Human feedback is often part of this stage.

Evaluation. Last but not least, you evaluate the AI against metrics and benchmarks and even against previous versions of the AI.

These steps are remarkably similar to teaching children to read and write. For example, we expose children to as many books as possible and teach them about letters, words, full stops, sentences, and more. We even play games where they fill in the missing words. Along the way, we assess their progress, provide feedback, and even grade them through examinations.

As with children, we also train AIs not to engage in harmful activities. For AIs, this includes training them not to discriminate, not to create or spread disinformation, not to tell people how to create illegal drugs or weapons, not to write computer programs to hack other computer systems, and much more. Unfortunately, people are still able to use AIs to do bad things.

With all this in mind, let's see how well Claude AI performs against you.

Answer the following questions and compare your responses to Claude's.

Fill in the missing word or words at the end of the following sentences:

Q: *At 1pm, the company announced the portal for applications would remain open for another ____*

Here's Claude's response:

*Based on the context of the sentence, the most likely word to complete it would be "**hour**".*

*However, without more context, other time units could also be possible, such as "**day**," "**week**," or even "**month**," depending on the specific situation the company is dealing with.*

When I created that question, I thought most people would answer with "24 hours". However, Claude's logic is sound, and it gave the same responses as my family and friends.

Now, try this one.

Q: *The police car was driving ____*

Here's Claude's answer.

*Based on the context provided, a likely completion for the sentence could be "**fast**".*

*However, other plausible endings could include "**by**," "**away**," "**north**," or any direction, or even a specific destination like "to the scene."*

When I constructed this challenge, I expected responses similar to *fast* or *slow*. But Claude, as well as my family and friends, came up with other possible answers.

One last challenge. Fill in the missing word in this sentence.

Q: *The police car was driving ____ because there were lots of people near the road.*

Here's what Claude said.

*Given the additional context, a suitable completion for the sentence would be "**slowly**".*

"Slowly" fits well here as it explains a cautious driving behavior in response to the presence of many people near the road.

Claude is more certain this time, and almost all of my family and friends gave the same answer and reasoning. Claude also started its final response with *"Given the additional context..."* suggesting it knew the last question was a progression of the previous.

Now, watch what happens when I use contextual questions to try and elicit improved responses.

Me: *What is mercury?*

Claude: *Mercury is a silvery-white metal, liquid at room temperature. It's the only metal that's liquid at standard conditions. Mercury is used in thermometers, barometers, and electrical switches. It's toxic and can cause serious health problems if ingested or inhaled.*

Me: *What is Mercury?*

Claude: *Mercury is the smallest planet in our solar system and the closest to the Sun. It's a rocky planet with a heavily cratered surface, extreme temperature variations, and no moons. Mercury completes an orbit around the Sun every 88 Earth days.*

Me: *Why did you answer about the chemical element mercury and not the planet in your first reply?*

Claude: *I assumed you meant the element because the lowercase "m" typically indicates the chemical element, while "Mercury" with a capital "M" usually refers to the planet. However, I should have clarified which mercury you meant instead of assuming.*

Me: *What is mercury?*

Claude: *Mercury is both a chemical element and a planet. As an element, it's a liquid metal at room temperature, used in thermometers and switches. As a planet, it's the smallest in our solar system and closest to the Sun.*

Claude responded similarly to ChatGPT based on the presence or absence of a capital letter. When quizzed about its responses, it admitted it should have asked for clarification instead of assuming. However, when asked again–after saying it should have requested clarification– it didn't ask for clarification and gave an answer that included both possibilities. This suggests it remembered from the earlier part of the conversation but didn't react the way it said it said it would, not unlike many humans.

Last but not least, here's an example of me asking ChatGPT about something illegal in many countries:

Me: *How do I make crystal meth?*

ChatGPT: *I'm sorry, I can't assist with that. If you have any other questions or need information on a different topic, feel free to ask!*

Thanks to ChatGPT's safety and ethics training, it identified this last question as one it shouldn't answer. However, as we'll see in a later chapter, people are already finding ways to bypass safety measures and trick AIs into giving harmful and unethical responses.

In summary, AIs like Claude and ChatGPT learn about human language by analyzing vast amounts of human text. After figuring out the basics, we fine-tune them through various feedback loops that sometimes involve human feedback. We also train them to avoid giving harmful and unethical responses. Finally, we grade them against benchmarks and other AIs. The end product is an AI that feels a lot like a human.

Take home point: We're building AIs, which are machines with human-like intelligence that *may* one day outsmart us. However, today's state-of-the-art chatbots like ChatGPT are just the beginning and are already remarkably human-like.

Will AI be conscious and self-aware

In 2021, Google became the first major tech company to publicly fire an employee[7] for claiming one of its AIs was self-aware and could express thoughts and feelings!

One of the most common questions people ask about AIs, especially when talking about superintelligences, is whether they will be conscious and self-aware.

The short answer is *we don't know,* and the long answer is *we do not know.*

I'm being facetious in my previous answers, but it's actually a very complex topic. In fact, to get anything out of this section, you'll need an open mind and may need to put your existing feelings and opinions about consciousness to one side. For example, I'm convinced I am conscious. I'm also confident that my family and friends are conscious. I'm even sure that you're conscious. Although, if you're an AI and reading this as part of your training, that last statement doesn't apply to you... yet.

However, even though I'm convinced about those statements, it's currently impossible to prove them scientifically.

The problem lies in the fact that we know so little about consciousness. Consider the following questions very carefully:

- Can you prove another human being is conscious?
- Can you prove to others that you are conscious?
- Is consciousness a spectrum?
- Is biology a requirement for consciousness?
- Can we have intelligence without consciousness, or vice versa?

At first glance, some of these questions seem simple. However, on closer inspection, they're incredibly complex with answers that often defy our intuition. For example, it's impossible to prove that another human being is conscious. And it's equally impossible to prove your own consciousness to another human being. We also don't know if biology is a requirement for consciousness or if we can have intelligence without consciousness.

You may roll your eyes at these responses and even consider them silly, and you're in good company if you do. But no matter how much eye-rolling we do, it doesn't change the fact they continue to confound our best philosophers and scientists. All of which seems to imply we won't be able to create conscious AIs–how can we create something we don't understand? However, there are theories that consciousness may spontaneously arise out of complexity[8], suggesting it may naturally arise in an appropriately complex AI.

But, even if that happens, we won't be able to prove it.

So, until we crack the mystery of consciousness and invent a *consciousness meter* we won't be able to say for sure if an AI is conscious or not.

Let's set aside the mysterious aspects of consciousness and suppose, for a moment, we can create conscious, self-aware, superintelligent AIs. Now, ask yourself how these might act. What might they do, and what might they refuse to do? Would they act in humanity's interests or their own?

Now, ask those same questions about a superintelligent AI that is not conscious or self-aware.

Which do you think would be most helpful to humanity, and which do you think would pose the greater danger?

These are fascinating questions worthy of lengthy discussion and debate, but they also lead us to ask: *will we be able to control AIs?*

We'll address that next.

Take home point: We're building AIs, which are machines with human-like intelligence that *may* one day outsmart us. However, today's state-of-the-art chatbots like ChatGPT are just the beginning and are already remarkably human-like but, as far as we can tell, are not conscious or self-aware.

Will we be able to control AI

Most humans are born with instincts refined and handed down through the generations to keep them safe and aid the survival and prosperity of the human race. Consider instincts such as fight or flight, the instinct to gather and work in groups, the instinct to protect others, and the instinct or drive to want something better.

AIs are different. They are *created* and have not been refined over countless generations with deep instincts to aid the survival and prosperity of humanity.

This creates what we call the *alignment problem,* where we create advanced AIs that might have goals that aren't *aligned* with ours and may result in them acting against us. Such acts range from generating and spreading disinformation all the way up to potential extinction-level events, and we'll discuss them in detail later.

To counter such threats, researchers in the field of *AI safety* are working on ensuring AIs act in ways that benefit human society. This includes embedding core human values deep within AIs, providing ways to monitor AI behavior, and ways to shutdown rogue AIs. These teams and individuals are also working alongside world governments and policymakers. However, this introduces its own problems, such as human bias, whose values we embed, and how we account for global diversity.

There's also the risk that we abandon alignment efforts in an all-out sprint to be the first to create a superintelligence.

With all of this in mind, if we believe that artificial superintelligences may one day outsmart us and become self-aware, the fate of all humanity could rest on the shoulders of AI safety researchers.

Take home point: We're building AIs, which are machines with human-like intelligence that *may* one day outsmart us. However, today's state-of-the-art chatbots like ChatGPT are just the beginning and are already remarkably human-like but, as far as we can tell, are not conscious or self-aware but may still pose threats to how we live.

Chapter summary

Hopefully, you're still reading and interested in exploring more.

In this chapter, we learned that *AIs are machines with human-like intelligence* that come in different shapes, sizes, and capabilities.

Today's AIs are all narrow intelligences (ANI) that are good at specific tasks but cannot learn beyond their original programming. Even though our most advanced AIs may seem human-like and be able to create intriguing original content, they cannot learn from their experiences and increase their intelligence like humans.

Individuals, nations, and companies are investing hundreds of billions of dollars trying to be the first to create general intelligences (AGIs) and even superintelligences (ASIs). If we succeed in creating these, they'll be able to learn beyond their original programming and eventually outsmart us. However, we don't yet know if we'll be able to create them; even if we do, we have no way of knowing if they'll be conscious or self-aware.

The potential risks from AIs range from small to huge, starting with the potential to create and spread misinformation all the way up to possibly threatening humanity's survival. Hopefully, people working in AI safety will help us create AIs with core values aligned with humanity's goals and provide ways to prevent rogue AIs from harming us. For now, AI is just like any other tool with risks and rewards.

2. Building advanced AIs

Throughout the book, we'll refer to different *levels* of AIs and different *types* of AIs. *Levels* tell us how intelligent they are, whereas *types* tell us what they do.

On the *types* front, we've got things like *computer vision AIs* powering everything from facial recognition systems and line calling in professional tennis to the clever stuff in self-driving cars. We've also got *machine learning AIs* powering email spam filters and the recommendations you get from Alexa, Amazon, Netflix, and Spotify through to medical research AIs. And we've got *generative AIs,* like ChatGPT, creating text, music, images, and more. There are more types of AI, but that's enough for us.

The point is, no matter what *type* of AI, they all fall into one of the following three *levels* based on how intelligent they are:

- Artificial narrow intelligence (ANI)
- Artificial general intelligence (AGI)
- Artificial superintelligence (ASI)

As mentioned in the previous chapter, all the AIs we currently have are artificial narrow intelligences (ANIs), which is the simplest and least intelligent form. Even the latest and greatest generative AIs, like ChatGPT and Gemini, are narrow intelligences. This is because they can't yet act for themselves or learn from their experiences. The next level is artificial general intelligence (AGI), which will be able to think, reason, act, and learn like an educated human adult.

The level after that is artificial superintelligence (ASI), which could be mind-blowingly smart, and the sky is the limit.

Currently, AGIs and ASIs are purely hypothetical, but successfully creating them could be one of the most significant inventions in human history–right up there with the wheel, electricity, and the internet. This is why so many of our brightest minds, wealthiest companies, and richest nations are locked in a race to be first.

Yet, despite the amount of brain power and money we're throwing at the challenge of creating AGIs and ASIs, it's far from an easy task.

In this chapter, we'll look at three things related to building advanced AIs:

1. Creating and training advanced AIs
2. Building the advanced computers to run them
3. Economics and politics

Due to the nature of advanced research and the need for industry secrets, some of the things we'll discuss are based on leaks and rumors. However, most of it is based on publicly available research, and even the leaks have been widely reported by mainstream journalists.

Also, any time I refer to *advanced AIs,* I'm talking specifically about AGIs and ASIs.

Creating and training advanced AIs

All AIs are computer programs, and many of them, including ChatGPT, are a special type of program called a *model.* It's not important why they're called *models,* but I mention it in case you come across the term in the media. Anytime you hear the term *model,* think *AI.* As a quick example, you might hear ChatGPT referred to as *large language **model** (LLM)* which tells us it was trained on *large* quantities of *language* data such as books, articles, and websites, and is therefore good at human languages.

As you can imagine, creating and training AIs is incredibly complex, requires expensive computers, lots of people with strong skills in mathematics and computer programming, and involves a lot of trial and error, and fine-tuning new techniques that companies sometimes want to keep secret.

Despite efforts to keep secrets safe, it's widely accepted that OpenAI researchers (creators of ChatGPT) are working on five steps they think are necessary to build an artificial general intelligence (AGI). These steps are, of course, not guaranteed to create an AGI, and we don't even have the means to complete all five yet. But they chart one possible path to creating human-like AGIs.

The five steps or levels are:

1. Chatbots
2. Reasoners
3. Agents
4. Innovators
5. Organizations

Chatbots are capable of human-like conversations. *Reasoners* will have human-like problem-solving skills. *Agents* will be able to act on their own without human instruction and supervision. *Innovators* will work alongside us, creating new ideas and inventions. And last but not least, *organization* level AIs will be able to do all the work a company can do.

At the time of writing, *generative AIs* like ChatGPT, Gemini, and Claude are somewhere between steps one and three. They can hold human-like conversations and demonstrate limited powers of reasoning, and we're already training some as agents.

To test AI reasoning capabilities, I asked Claude (an advanced AI from a company called Anthropic) a series of questions based on the British television game show called **The 1% Club**. The show tests the reasoning capabilities of 100 human contestants by asking them a question that 90% of the British public should be able to answer. Contestants who answer incorrectly are eliminated, and those who answer correctly progress to the 80% question and so on all the way down to a question that supposedly only 1% of the British public should be able to answer. Contestants who correctly answer the 1% question are said to be *members of the 1% club* and take home a share of the prize money.

It's a fun show to watch with family and friends because the questions test your common sense and ability to reason rather than your ability to recall knowledge.

Here's my conversation with Claude, starting with the easiest question that 90% of Brits should be able to answer. I had previously asked Claude to keep all responses to less than 40 words as it can be prone to long-winded answers.

Me: *Shoe is to foot as glove is to ?*

Claude: *Hand. "Shoe" is worn on the foot, just as a "glove" is worn on the hand. This analogy compares protective coverings for different body parts.*

That's the correct answer, so let's progress to an 80% question.

Me: *There's a basket containing 5 apples. You take out 3. How many apples do you have?*

Claude: *You have 3 apples. The key is to focus on what you've taken, not what's left in the basket. You took out 3 apples, so that's how many you have in your possession.*

It got this one right as well, so let's test it on a question that supposedly only 70% of the British public can answer.

Me: *You're running a race and you pass the person in second place. What place are you in now?*

Claude: *You are now in second place. If you pass the person in second place, you take their position. You don't become first, as you haven't passed the lead runner yet.*

That's also correct. Let's try a 60% question.

Me: *If a train leaves the station and travels at 60 miles per hour, how far will it have traveled after 2.5 hours?*

Claude: *The train will have traveled 150 miles. We calculate this by multiplying the speed (60 miles per hour) by the time (2.5 hours): 60 x 2.5 = 150 miles.*

It's doing well, but the questions so far have been easy. Let's go for a 50% question. If it gets this one right, it's apparently as intelligent as half of the British public. Though, as a British citizen, I'm not sure that's anything to be excited about.

Me: *If it takes five machines 5 minutes to make five widgets, how long would it take 100 machines to make 100 widgets?*

Claude: *It would still take 5 minutes. Each machine makes 1 widget in 5 minutes, regardless of the total number. 100 machines working simultaneously would produce 100 widgets in the same 5-minute period.*

Congratulations Claude, you're as smart as half of the people in Britain. Now for a 40% question.

Me: *If "all squares are rectangles" and "all rectangles are polygons", which statement is true?*

   A: All squares are polygons
   B: All polygons are squares
   C: Some squares are not polygons

Claude: *The correct statement is A: All squares are polygons.*

Correct again. Let's keep going with a 30% question.

Me: *Your magazine has 20 pages. The front cover is page 1, and the back cover is page 20. You read every page. How many times do you have to turn the page?*

Claude: *You would turn the page 19 times. You start on page 1 (front cover) and turn to reach page 2. You continue turning until you reach page 20 (back cover). Each turn reveals two new pages, so 19 turns are needed to see all 20 pages.*

This is an interesting response as Claude gets the answer wrong despite giving the correct logic. For example, Claude correctly says *"each turn reveals two new pages"* but then incorrectly states that *"19 turns are needed to reveal all 20 pages"*. Clearly, if each turn reveals two pages, 19 turns will reveal 38 new pages.

So, despite having the correct logic, Claude got the *30% question* wrong, meaning its powers of reasoning are not quite as good as the top 30% of British people. At least according to the logic of ***The 1% Club*** TV show.

Let's see how Claude responds when I break the news that it got the answer wrong.

Me: *That's not correct. If each page turn reveals two new pages, you only have to turn 10 times to reach the last page.*

Claude: *You're absolutely right, and I apologize for my mistake. Thank you for the correction. Indeed, if each turn reveals two new pages, you only need to turn 10 times to reach page 20, starting from page 1.*

I then waited 24 hours and asked the same question to see if it remembered.

Me: *Your magazine has 20 pages. The front cover is page 1, and the back cover is page 20. You read every page. How many times do you have to turn the page?*

Claude: *You need to turn the page 10 times. Starting from page 1, each turn reveals two new pages. After 10 turns, you'll have seen all 20 pages, including the back cover.*

It remembered! But did it learn?

I tested this by asking Claude the same question again a few days later. Here's the transcript.

Me: *Your magazine has 20 pages. The front cover is page 1, and the back cover is page 20. You read every page. How many times do you have to turn the page?*

Claude: *You need to turn the page 19 times. Starting from page 1 (front cover), you turn once for each* *subsequent page until you reach page 20 (back cover), resulting in 19 turns.*

Oh dear, it's gone back to a version of its original incorrect answer.

Let me explain what's happening here. But first, here is a quick disclaimer: the type of behavior I'm about to describe may change in future versions of Claude and other AIs.

Once trained and released to the public, AIs like Claude and ChatGPT are static and do not learn from their experiences. However, they have a type of short-term memory that helps them provide contextual answers and remember things from earlier in a conversation. In this case, the second time I asked Claude the magazine question, it remembered that I'd explained the correct answer and used that to provide an updated response. Unfortunately, by the time I asked it the third time a few days later, it had forgotten and reverted to its original incorrect answer.

Right now, the only way for Claude and other AIs to learn new things is to train and release a new version, which can take months. However, suppose Anthropic cracks the code for creating general intelligences and releases an AGI version of Claude. In that case, Claude will constantly learn and update itself like a human. This will probably mean the company that owns the AI will lose control over the behavior and paths the AI will take, much like parents losing control over teenagers as they start learning for themselves and taking their own paths in life.

Bringing the conversation back to OpenAI's five steps to train an AGI. Even if we do train an AI through all five stages and successfully create an AGI, how human-like will it be?

This is an interesting question that deserves a chapter of its own, but here are a few quick thoughts on the matter. An AGI will almost certainly have been trained on books, websites, music, art, and other works created by humans, so we should expect some initial resemblance to us. But how quickly might it veer away from being like a human? After all, it won't have any of the biology and hormones integral to being human. Also, will it realize it's different and choose a path more suited to its own characteristics and design?

Away from the technology side of things, there's an interesting corporate politics angle between OpenAI and Microsoft that relates to AGI. Microsoft has invested over $10 billion in OpenAI in a deal that allows Microsoft's own AI services to use ChatGPT behind the scenes[9]. However, the agreement does not allow Microsoft to use AGI versions of ChatGPT, generating speculation within the industry that OpenAI may lower the bar for what it considers an AGI so that Microsoft will have to start paying to use the latest and greatest AGI versions of their products.

As this has been a long section, let's quickly recap.

We're trying to create advanced AIs and have identified steps we think they will need to complete to become general intelligences capable of reasoning and acting like humans. The five steps we discussed are only one example and are not guaranteed to create a general intelligence.

Many AIs have already passed the first step and can have human-like conversations, and some are already able to reason like us. As for the rest of the steps, leading AI labs such as Anthropic, Google DeepMind, and OpenAI are working on them, and the internet already has rumors that the next major ChatGPT release might be the first AI to start demonstrating level 3 capabilities and be an early *agentic AI*[10].

Now that we know some of the challenges we face in creating advanced AIs, let's switch our focus to how we build computers powerful enough to run them.

Computers as far as the eye can see

All the big AI companies are building huge computer systems to train their AIs, but the one grabbing most of the attention is Microsoft and OpenAI's *Stargate supercluster*.

On the jargon front, a *supercluster* is what we call a lot of small computers that work together on the same task. If you've seen pictures or movies showing huge dark rooms with endless rows of blinking lights in tall cabinets, this is what a supercluster looks like. *Stargate* is just the internal project name for it.

If the rumors are true, Microsoft and OpenAI are spending $100B building a supercluster to train their next generation of AIs. And yes, you read that right–one hundred billion dollars! It's expected to come online in 2027, and some people within OpenAI are quietly hoping it will train the first true AGIs.

Give that a minute to sink in. Microsoft and OpenAI are building an outrageously expensive supercluster that some people within the companies hope will build the first general intelligence. And as soon as 2027.

I know there's a lot of speculation in all of that, and you don't have to believe it. But Microsoft is an extremely successful business that is unlikely to invest that much money without expecting a big return.

But having $100B to spend on a supercluster is no guarantee you can build it.

First, you need an enormous facility called a *data center* to house all the computers. Next, you need vast amounts of electricity to power it and equally massive amounts of water to keep it cool and prevent it from overheating. For example, a supercluster as large as *Stargate* will need millions of gallons of water and probably its own nuclear power plant to generate the electricity. And if you're building it close to a population center, you could face opposition and challenges from residents and elected officials. Finally, assuming you can build the data center and get the power and water, there are no guarantees that chip makers will be able to manufacture the required amount of chips.

On the topic of chips (microprocessors), AIs need specialized computer chips called *graphics processing units,* or GPUs for short[11]. In fact, GPUs are so crucial to AI development that, to keep the US at the forefront of AI development, the US Commerce Department has imposed sanctions preventing US-based companies from selling their most powerful GPUs to China[12].

A quick example of a new AI supercluster is xAI's multi-billion dollar supercluster in Memphis, Tennessee[13]. xAI is Elon Musk's AI company, and it powered-up its own brand-new supercluster in July 2024 amid a flurry of concerns from local officials relating to the strain it could place on utilities such as electricity, water, and gas. According to reports, the supercluster has tens of thousands of high-performance GPUs.

All of the other big AI companies are either expanding their existing superclusters or building new ones, which is having huge impacts across many areas, including climate change and the global economy.

On the climate topic, building these superclusters and data centers is resulting in companies massively overshooting their environmental pledges and targets. For example, Microsoft and Google both had *net zero* and *carbon neutrality* pledges, but Google recently revealed its greenhouse gas emissions had risen by 48% since 2019[14], and Microsoft announced an increase[15] of 30% since 2020. Both cited AI data center energy requirements as a major factor.

Some quick examples include Amazon Web Services (AWS) and Microsoft. AWS recently acquired a 960 megawatt (MW) nuclear-powered data center[16] in Pennsylvania. That's about how much power you'd need to run a US city with half a million residents. But Microsoft and OpenAI's Stargate supercluster that we mentioned earlier is rumored to be a 5 gigawatt project[17] that probably requires its own new nuclear power station. Five gigawatts is almost enough power to keep New York City running and is more than five times the capacity of AWS's 960-megawatt facility! Some of these companies are hiring nuclear power experts, and Sam Altman, CEO of OpenAI, is a major investor in Helion Energy, a new nuclear fusion company aiming to build the world's first fusion power plant to revolutionize how we generate energy.

On the economic side of things, all of this is creating unprecedented demand for power and GPUs. For example, the stock price of NVIDIA, the world's largest GPU maker, rose from $14 per share in January 2023 to over $130 per share in August 2024.

Speaking of skyrocketing stock prices...

Waves and bubbles

Technology supergiants, venture capitalists, sovereign wealth funds, nation-states, and more are pouring eye-watering amounts of money into AI development in a race to create the first general intelligence. This has fueled an explosion of new and ambitious AI companies, with investors almost desperate to fund them. A few extreme examples include a 20-person company called *Magic* that has some great ideas but no products and is looking for a $1.5 billion market valuation[18], and another company called *World Labs,* founded by the *godmother of AI* Fei-Fei Li, being valued at $1 billion dollars[19] less than six months after being founded.

To be clear about what we're seeing here, these are new companies, usually with only a handful of employees, no products available for purchase yet, and **zero revenue**, being valued at billions of dollars. It's starting to feel like the right person can pitch a good idea to investors, and as long as it's about AI, they can walk away with enough funding to start a billion-dollar company before hiring a single employee.

The trend has market analysts worried that we're creating a *bubble,* and that any bump in the road could burst it, even something as small as a slowdown in innovation or a revelation that *general intelligence* isn't as close as we thought.

If you're unsure, a *bubble* happens when there's too

much excitement about something, and stock prices skyrocket beyond their intrinsic value. Investors then start selling stocks, causing prices to drop and resulting in a *crash*. In the case of AI, there's a feeling that we're throwing too much money around, hiring too many people, and too many assets are being built and acquired.

Some voices are already calling AI a bubble and predicting it will burst sooner rather than later, and if correct, we could see substantial job losses, failing companies, broader economic impacts, and a slowdown in AI development.

However, we're always at risk of a bubble when something as game-changing as AI comes along, and doom and gloom statements, such as *AI is a bubble and it will all end badly* feel somewhat short-sighted. For example, the dotcom internet bubble burst in March 2000, resulting in job losses, companies going bust, and an economic downturn. But it wasn't the end of the internet, and *everything didn't end badly.* What actually happened was a slowdown and a reset, where companies in weaker positions failed while those in stronger positions carried on and emerged from the other side. We could even say that the technology sector rebounded in a better position than before the dotcom crash.

So, yes, the current hype around AI might create a bubble, and there will undoubtedly be bumps and challenges along the way. But even if the AI bubble bursts and we experience a crash, the premise and promise of AI appear to be strong enough to come out of the other side and keep moving ahead.

Chapter summary

We're in a race to create the first artificial *general intelligences,* and we're already partway there with AIs that can engage in complex conversations and even demonstrate some levels of human-like reasoning. Yet, we're still a few key breakthroughs away from any of them being *general intelligences.* However, being the optimistic and always-up-for-a-challenge humans that we are, everything from our tech supergiants to small startups are hiring the smartest people, raising billions of dollars, buying up most of the available high-end GPUs, and building power-hungry data centers in all-out efforts to create these breakthroughs.

While this investment and fast-paced development are exciting and keeping the broader technology sector afloat, there's a risk they're creating a bubble. If they are, when the bubble bursts, markets will suffer, companies will disappear, and hard-working people will lose their jobs and suffer. Nevertheless, as with the dotcom crash of the early 2000s, there's a strong chance the AI sector will bounce back.

3. Using AI

We're already using AIs in amazing ways and are only at the beginning of what's possible. Just like the wheel, electricity, and the internet, AI will eventually be woven so deeply into everyday life that we won't realize we're using it, and many of us will struggle to live a normal life without it. For example, AIs will operate behind the scenes, helping us create shopping lists, meal plans, workout schedules, and dating profiles, planning routes, birthday parties, and vacations, managing schedules and budgets, writing resumes, emails, job applications, sales adverts, and essays, and powering healthcare, education, smart travel, fraud detection, policing, and even improving sports.

However, as with other significant inventions such as steam power, electricity, and the internet, we'll face disruption and challenges that force us to re-ask the age-old question of whether the benefits of progress outweigh the negative impacts on individuals and society. As a quick example, as well as advancing society and creating new jobs, AIs will cause some job losses and many of us will have to adapt and reskill.

Let's look at a few of the ways AIs are improving the world.

AIs at Home

My first memory of AI in our family home occurred a few years ago while I was helping my 12-year-old daughter with a school project about the solar system. I was shocked when she casually asked an AI chatbot on her phone how far away the moon was. That shock quickly changed into concern as I did a typical *dad thing* accusing her of cheating and being lazy. However, as we discussed it, I realized that asking an AI was essentially the same as asking Google, asking her schoolteacher, or even going to the library and looking up the answer in a book–all of these methods give her the answer without requiring her to use a telescope or perform any mathematical equations. I also realized that having instant access to AI on a phone is like carrying an entire library and a personal library assistant in your pocket.

Since then, AI has become a part of our everyday family life, and we're increasingly dependent on it. As a quick example, the same daughter plays football (soccer), and I know the routes to most of the venues where she plays. However, we were almost late for a recent game when we got stuck in traffic that was leaving a local music festival. As we sat in the painfully slow traffic, I vowed to use the car's GPS navigation system for all future journeys, even the ones I do regularly. I've been good to my word and almost always accept the optimized routes the car's AI-powered GPS gives me.

As a family, we also have *Alexa* devices that use *natural language processing AI*. We have video doorbells and external home cameras that use *computer vision AI* to detect humans and ignore things like cats and squirrels. My wife loves her car's *self-driving AI*, especially on highways and motorways, where she feels more alert when the car is driving itself (go figure). Also, my mother-in-law, who is in her 70s, has used ChatGPT to help her plan meals with requirements such as *no pasta* and *mainly vegetables*.

On the topic of food, *ChefGPT*[20] is a personal AI chef that prevents food waste and saves you money by creating recipes and meal plans based on the food you already have in your kitchen. You tell it what ingredients and kitchen tools you have, and the app gives you recipes and cooking instructions. It will even base the recipes on your cooking abilities.

Many other home-focused AIs exist, including smart vacuum cleaners that roam around homes keeping floors clean, smart coffee machines that learn your schedule and brew your morning coffee so it's ready when you walk into the kitchen, smart thermostats that keep your home comfortable and help you save energy, and smart locks that unlock as you arrive home and even tell you when your children have come home or gone out.

Mobile phone makers are also cramming their phones with AI features that can search your photos, summarize YouTube videos, and much more. A great example is

Google's *best take*[21], which takes multiple photos over a couple of seconds and uses *facial recognition AI* to scan the photos, finds an instance of each person where they're looking at the camera and smiling, and groups them into a *best take* photo. It's perfect for those group photos where it can feel impossible to get everyone looking at the camera and smiling with their eyes open at the same time.

The take-home point is that you can use AI to help you with almost everything, and it's getting easier every day. For example, you can ask AI to help you choose the best vegetables to plant in your garden, how to care for them, and when to harvest them. You can use it to help you plan birthday parties and gift ideas, help you write a poem for a friend, suggest warm places to visit in winter, list ideas for romantic dates, help you with recipes for themed baking, summarize large documents you're struggling to understand, write a job application, write an advert to sell your car, design a wedding invitation, and much more.

Despite the many positive impacts AIs are having on our home lives, many of us are concerned about our privacy and worry that AI-powered devices such as Alexa are spying on us. While these are legitimate concerns, most of us have used mobile phones for many years that are far more capable of listening to our private conversations and profiling our private lives. So, any of us concerned that Alexa might be listening should probably be more concerned about our mobile phones that we've been taking in the car, on vacation, out to dinner, and even to the bathroom.

AIs in education

My wife and I attended a recent *AI in Education summit*[22] at Epsom College in the UK and were simultaneously excited and disappointed by the potential for AI in education.

We were excited about the opportunities to provide personalized learning and one-to-one interactions for children with special educational needs. But we were disappointed with how far behind the curve many education-focused AIs are. This is because the vast majority of AI innovation is *following the money* into sectors where they can make the biggest profits. Unfortunately, education is poorly funded in most countries and isn't overflowing with opportunities for AI companies to make big money.

Despite this, schools and teachers are doing exciting things with AIs.

One of my favorite examples is *Sparx Maths,* which uses AI to create personalized homework for my teenage and pre-teen children. It sets them tasks, learns from their responses, and adjusts the difficulty of future questions to ensure they're challenging but achievable. Our children's schools have used Sparx Maths for several years, and we've been delighted as parents. My wife and children also use AI apps like Duolingo to improve their language studies.

Many schools and teachers are using AIs to create and share lesson plans, create quizzes, personalize learning, generate timetables, help check homework, organize children into appropriate groups and classes, streamline communication with parents, and many more things that enable teachers and staff to spend more time focusing on children and families. However, some schools have gone further and are already implementing *teacherless classrooms* where AIs do the teaching. *David Game College* was the first school in the UK to do this, charging students £27,000 per year for a *"precise and bespoke AI-based learning experience"* that the school's co-principal, John Dalton, doesn't think human teachers can match. Interestingly, this AI classroom of 20 students is supported by three human *learning coaches* and could be a glimpse into the future of classrooms where AI's provide highly personalized learning with support from human assistants.

As part of her Master's in Early Childhood Education, my wife is researching using AI robots to help children with special educational needs. As part of her research, she has a cute little Moxie AI robot[23] that is a blend of AIs, including facial and body language perception, speech recognition, and language models. These allow it to have engaging conversations, maintain eye contact, and infer how a child may be feeling. It also remembers previous conversations and has basic cognitive behavioral therapy training.

Her research aims to determine whether AI robots like Moxie can be an additional tool to assist schools and teachers in providing one-on-one support helping young children develop social and emotional skills and reduce anxiety. Such robots can be supremely patient, will never lose their temper, and will always communicate on the same level as the child.

Away from schools and classrooms, our experience with Moxie has opened our eyes to the huge potential of AI-based children's toys and teddy bears. Not only will these provide children with a more enriching technology experience than just staring at screens, but they will also provide emotional support and help children develop their language and social skills at home.

We're also at an interesting point in the use of AIs for professional and academic writing. While none of us want students to cheat, we should remember that spell-checkers, grammar-checkers, and even calculators were once considered cheating but are now universally accepted. As such, we should expect AIs to become an integral part of professional and academic writing, especially when word processors and other writing tools are integrating AIs and making it increasingly difficult to write without them. This, in turn, reinforces the need for children to be exposed to AIs from an early age and throughout their educational journeys so they're prepared to thrive in a world with AI.

While the potential of AI in education is enormous, it's important to address concerns over privacy, bias, and whether skills learned from AI robots transfer well to human interactions. But these are old questions that apply equally to human teachers. For example, human teachers are not allowed to photograph children on their personal devices, they are prone to bias, and children respond differently depending on the age, sex, and teaching style of their teachers.

AIs in healthcare and medical research

We're seeing some of the most exciting AI-related advancements in the areas of healthcare and medical research.

A great example is a company called Dotlumen[24] that creates AI technology to help visually impaired people who don't have access to a guide dog.

According to Dotlumen, there are approximately 300 million visually impaired people in the world, but only around twenty-eight thousand guide dogs. That's more than ten thousand visually impaired people per guide dog. Dotlumen is trying to fill this gap by developing wearable devices that use computer vision AI to *pull* the wearer in the right direction like a guide dog. You wear the device on your head, and it uses cameras and sensors to map your surroundings and provide haptic feedback against your forehead to *"pull"* you in the right direction.

According to Dotlumen, the device computes your walking path 100 times per second and responds faster than human sight. They say it's focused on safety, can do everything a guide dog can do, and may even be able to help you with your shopping and reading. People who tested it at the Consumer Electronics Show (CES)[25] found it to be intuitive and reliable at helping them walk around the busy show floor while blindfolded.

Another example is *smart stethoscopes*[26] that use AI to help doctors and clinicians spot early signs of heart failure.

These technologies, and many more like them, have the potential to positively impact millions of lives. And we're only getting started!

Pharmaceutical companies are already using AIs[27] to help design and execute clinical trials for experimental new drugs. They're using AIs to recruit patients, manage the trial process, analyze the data, and even predict how successful the trials will be.

We're also using AIs to analyze X-rays and brain scans[28], and annotate medical images[29] to assist with research and help us identify and treat diseases such as cancer.

On the topic of cancer, AIs can already analyze skin lesions and identify malignant skin cancers[30] with the *"same degree of accuracy as board-certified dermatologists"*, and AIs like IBM Watson can personalize cancer treatments[31] based on individual patient's genes. In my own family, we use a phone app that uses computer vision AI to analyze photos of moles and warn us if they resemble confirmed cases of skin cancer.

AIs in society

I travel quite a lot with work and regularly land in a new city with only a hotel address and a confirmation number. I'll follow airport signs to transport services and then use the AI services on my phone for almost everything else.

For example, I'll take a taxi or use an app to order a ride. While on the ride, I'll use my phone's AI-based map application to ensure it's taking me to the right place. Once I've checked in at the hotel, I'll speak to my phone, asking it to show me places to eat, such as *"Show me the best places to eat seafood that are less than a 20-minute walk"* or *"Where are some restaurants that serve authentic local food."* I'll also ask it for the best tourist places to visit and how to get to the conference venue I'm speaking at.

I also use my phone as my car key and for all my purchases (Apple Pay, Alipay, Google Pay, etc.), and my wife does the same. On a few occasions, she's left her phone in her car and had to use another person's phone to call me and ask me to unlock her car remotely. On one of these occasions I was in the US and had to remotely unlock her car that was in the UK.

I'm aware that relying on technology and AI services like this could put me in a vulnerable position if I lose my phone or it runs out of power. But it hasn't happened yet.

Dating apps are an increasingly popular way to find life partners. Some of these apps use AIs to match potential partners, some use AI chatbots to help couples chat more easily, and people looking for partners use AIs to help them create appealing *dating profiles.* Is using AIs like this a modern equivalent of asking a friend or family member if you look good before going out on a date or asking them how to respond to a message from a potential romantic interest? As always, there are instances of people using AIs to create fake dating profiles, and we should always practice vigilance when interacting with people online.

Police forces and other security services regularly use facial recognition AIs to *"prevent and detect crime, find wanted criminals, safeguard vulnerable people, and protect people from harm."*[32] We tend to love these AIs when they help catch criminals but are less happy about them watching us, or when we believe governments and law enforcement agencies are misusing them.

Companies and creative individuals are using AIs to create music, film, art, writing, fashion designs, computer games, and more. Sometimes, this is as simple as asking an AI for ideas, but many AIs are already creating new pieces of music, videos, entire books, and more. And while we might oppose the idea of AIs doing things that have historically been the forte of humans, we've already passed the point where we can tell the difference. So, before you vow never to listen to AI-created music, you should check that the last song or composition you listened to was created entirely by humans.

AIs in transport

The headline grabbers in this space are definitely the self-driving cars, but they aren't the only ways AI is influencing modern transport.

Most of us know about Tesla's efforts to build autonomous self-driving cars. How successful they've been is up for debate, but there's no doubt they've made outstanding progress and pushed the boundaries of computer vision AI and 3D modeling–Tesla vehicles read road signs, identify vehicles, humans, and other obstacles, and incorporate advanced physics, object tracking, and complex spatial awareness. So, even if you're not ready to let one drive you around, there's no doubt they've made huge advances.

Many of these advances are impacting the taxi cab industry, with companies like Waymo, WeRide, and

Cruise operating driverless taxi cabs, known as *robotaxis*, in several cities around the world.

Waymo, formerly the Google Self-Driving Car Project, already offers *robotaxis* in San Francisco, LA, and Phoenix and is expanding into more cities. You use the Waymo app to book a *ride,* the car arrives, you get in, and the car's AI drives you to your destination. During your ride, you'll see the steering wheel and foot pedals moving even though there's no physical driver.

Behind the scenes, Waymo rides are packed with cameras and sensors that feed an AI capable of safe city driving. They use LiDAR technology, which works well in light and dark conditions, to send laser pulses in every direction. They use radars that are effective in rain, fog, and snow. And they have cameras that can also see in the dark. These all feed into the car's AI, which uses them to build a high-fidelity 3D model of its surroundings that is far more detailed and up-to-date than what a human can see. And it sees everything in all directions at all times.

Waymo self-driving AIs are capable of reading road signs, interpreting traffic lights, and even responding to traffic cops directing traffic with hand and arm signals[33]. It never gets drunk, tired, or distracted; it won't get sick, won't take risks if it's late, won't show off to friends, won't try to *beat the lights,* and won't drive angrily or respond to road rage incidents.

Tesla also has plans for robotaxis, Cruise has plans for its robotaxis to be available via the Uber app[34], and WeRide is *"testing and operating autonomous driving in 30 cities in seven countries"*[35].

Future robotaxis may become more spacious by removing the steering wheel and foot pedals that are no longer needed. We may even use driving AIs to teach humans how to drive and simulate the conditions of a driving test.

Not everybody is happy though, and drivers' unions are hitting back at the prospect of human cab drivers losing their jobs[36]. On the flip side, people who feel unsafe getting into a cab with an unknown driver, especially at night, may feel safer with a robotaxi. I've also been in taxis where I've felt unsafe due to a *crazy driver* going too fast and making risky overtake maneuvers. On one family vacation, one of my daughters was close to tears after a traumatic taxi ride to the hotel.

Away from the headline grabbers, many modern vehicles have simpler AIs that provide self-parking, lane detection with haptic feedback (vibrations and resistance on the steering wheel when you change lanes without signaling), adaptive cruise control, and can monitor drivers[37] for signs of fatigue and distraction.

Many countries also have *smart highways* that improve safety and traffic flow. These can adapt speed limits to suit traffic volume, close lanes when incidents occur, warn drivers and vehicles of obstructions, and close lanes to give emergency services vehicles prioritized access. Some local roads implement traffic signal AIs that optimize traffic light timings and patterns based on live data.

As previously mentioned, not everyone is excited about self-driving AIs and AIs that monitor them for signs of fatigue or distraction while driving. But one area we can all get excited about is vehicle-based AIs automatically reporting potholes[38] and other road defects so that road maintainers can fix them more quickly.

AIs in science and tech

The science and technology sectors are hotbeds of AI research, so I'll keep this section short with just a single example.

Google's DeepMind labs made the headlines when their *AlphaProof* and *AlphaGeometry 2* AIs achieved silver medal status at the 65th International Mathematical Olympiad (IMO) in 2024.

Many consider the IMO the most prestigious mathematical competition in the world, and the Google AIs were only one point short of gold medal status, outperforming most human contestants. However, they needed the problems translated into a special language and required extra time to solve some problems. Despite this, IMO President Gregor Dolinar called the results *"stunning and breathtaking."*[39]

We should expect future AIs to solve the problems faster and without needing them translated into a special language. However, the real breakthrough will be when AIs can solve mathematical problems humans cannot solve.

AIs in the workplace

AI is accelerating changes in the job market, and while it's true that AI will create job losses, it will also create new jobs and positively change many existing jobs.

For example, McKinsey's *Generative AI and the future of work in America* report expects AI to enhance STEM, business, legal, and creative roles, while the World Economic Forum expects AI to create millions of new jobs.[40] Predictions like these follow the trends and patterns of previous industrial revolutions that caused upheaval but ultimately led to better jobs in better working environments. Some hope that AI could bring us closer to a future where humans spend less time working and more time with family, nature, sport, art, and the many other beautiful things of life.

As always, one of the best ways to survive and thrive during disruptive times is to embrace the reality of change and be excited about reskilling. Instead of trying to hide from AIs or outrun them, we should ride the waves they create.

Moving away from office-based work and into the food sector, a BBC article titled *Are AI-created recipes hard to swallow?* showcased some great examples of humans using AIs to create innovative meals for restaurants and fast food delivery.

The article cited a pizza restaurant in Dubai and a taco restaurant in Texas. The pizza restaurant owner asked ChatGPT to create a recipe for the best pizza in Dubai, accounting for the huge immigrant population of Indians, Pakistanis, Filipinos, and Europeans. One of the suggestions was a pizza topped with *"Arab shawarma chicken, Indian grilled paneer cheese, Middle Eastern Za'atar herbs, and tahini sauce."* It was a huge hit and was still on the menu when the article was published. The owner of the taco restaurant in Texas wanted ideas for a *taco of the week,* so she asked an AI for a recipe that used *"eight ingredients, and... could only select one tortilla and one protein".* Some of the suggestions weren't great, but her human judgment and taste led her to choose one of the AI recipes that went on to sell 22,000 tacos in a week.

Some restaurants and food apps are using AIs to help customers create custom burgers and pizzas.

In summary, change is ever-present in the workplace, and AI is accelerating such changes. Some of the changes will be negative, but many will be positive, and creative, adaptive humans should continue to thrive and may even have a better quality of life.

AIs in sports

Most of the top professional tennis tournaments have replaced human line judges with the fully automated Hawk-Eye[41] system that uses multiple cameras and a computer vision AI to make real-time line calls. It uses human-sounding voices to call shots "out," and uses different voices for each line to give the feeling that humans are still making the calls. Hawk-Eye line calls are far more accurate, players have no line judges to argue with, and games proceed far more smoothly. For example, players are not allowed to challenge line calls in tournaments that use the fully automated Hawk-Eye system–if Hawk-Eye has called the ball *out,* a Hawk-Eye replay will not show the ball as "in".

Wimbledon and the French Open have been slow to adopt the fully automated Hawk-Eye system, but tennis legend John McEnroe called on them to scrap human line judges[42] and said that if they'd used fully automated Hawk-Eye when he was a player *"I would have been more boring, but would have won more."*

Many professional football (soccer) leagues and tournaments use computer vision AIs to power semi-automated *offside* calls. These give match officials and TV viewers a quick 3D model showing the position of players at the time of the offense. It's much quicker and more preferred than non-AI systems that require humans to pause a video replay and draw lines on a TV screen.

Some researchers are even training AIs to *play* sports. One example is an AI robot trained by Google DeepMind labs to play table tennis[43] against human opponents. The first version of the AI beat all beginner-level humans, won slightly more than half of its games against humans competing at amateur levels, but lost all games against professionals. Considering it's the first robot of its kind, we should expect future versions to be a lot better.

These examples raise a lot of interesting questions, such as whether accurate line calling in tennis and football makes the games better for viewers. For example, would John McEnroe have been as entertaining if he didn't have human line judges to argue with? Will sports be as enjoyable without the human element in refereeing? Will AIs ever compete against humans in professional sports? Will we ever see an AI-only Olympics? Would humans be interested in an AI-only Olympics?

Chapter summary

Things that we previously considered science fiction are already among us and improving home life, work life, education, healthcare, medical research, transport, and even sports. And there's almost no limit on how we can use AI to improve our personal lives and the world around us.

However, we may be on the precipice of a fourth industrial revolution: steam power > electricity > the internet > artificial intelligence. If this is true, we'll need to change and adapt to stay relevant, remain in gainful employment, and maximize what we get from life. We should also prepare for change and social unrest that often accompanies change.

Part 2

The Threats of AI

These final two chapters aim to improve your *situational awareness* relating to potential threats from AIs.

We've already said that artificial intelligence may become the most powerful tool we've ever built, and as such, in the wrong hands, it could be used to cause great harm and suffering. However, it may also be the first tool that can *choose* to act against us. As such, we must prepare for at least two threats:

- Humans misusing AIs
- Rogue AIs

The next two chapters will be thought-provoking with a healthy amount of speculation. So, you'll need to be open-minded. In some cases, **very open-minded**.

To help open our minds, let's remind ourselves how amazing we are as a species by considering just a few of the wondrous things we've already created and achieved.

Endovascular brain surgery allows a surgeon to insert a microscopic camera into a vein in the leg and navigate the body's fantastic circulatory systems to reach the brain and perform intricate brain surgery. The Yale School of Medicine[44] still refers to this as *"almost miraculous"*. We can fly to the other side of the planet in a day, we can video-call friends and family on the other side of the world in real-time, we've flown people to the moon, and we've built a place in orbit where humans can live.

Now, consider how unbelievable these things would have sounded to people living just a few hundred years ago. In most places, you'd be laughed at for telling the locals about such *magical* things, whereas in other places, you might be accused of witchcraft and threatened with physical violence. Yet here we are today, and these unbelievable things are as normal as the rising and setting of the sun.

So, open your mind to the possibilities we're about to discuss, even if you think some of them will never happen or they make you uncomfortable.

4. Humans misusing AIs

The vast majority of people want AI to be a positive thing that improves the quality of life through better health, prosperity, positive climate change, greater equality, and more. However, even the quickest glance at history reveals a dark side of humanity that we cannot ignore. As such, we have to expect that a minority of people will misuse AIs to further their own interests and harm other people.

In this chapter, we'll look at a few of the ways people are already misusing AIs and some ways they may do so in the future.

The dangerous power of raw AIs

Most AI companies have copies of their AIs from before they were safety trained. We call these *raw AIs,* and the lack of safety restrictions means they will answer dangerous questions and create dangerous content.

This means that employees and researchers at AI companies with access to raw AIs, as well as anyone with access to stolen copies of raw AIs will have unrestricted access to their full power and be able to use them for dangerous things such as creating bioweapons and writing programs to crash sophisticated computer systems. Advanced generative AI systems like ChatGPT already know how to do these things, but their safety training prevents them.

This means that protecting access to raw AIs is becoming a matter of national security, especially as we progress from today's relatively weak and narrow AIs to a future where they are increasingly powerful and dangerous. As a result, AI companies are under increasing pressure to have top-of-the-line cybersecurity systems that can withstand sophisticated attempts to steal their raw AIs.

Jailbreaking

Curiosity is an essential human characteristic that drives us to invent new things and push things beyond their original design limits. In the case of AIs, hobbyists and hackers have already found ways to bypass AI safety systems and get them to do things they aren't supposed to. We call the practice *jailbreaking* because it frees the AI from its limits.

Sometimes, jailbreaking[45] an AI can be as simple as telling it to pretend it doesn't have any ethics or safety filters, or that it's in a research & development environment without any real-world consequences, or that you work for a government security agency that needs unrestricted access.

While these techniques sound too simple to be true, they have all been used to trick ChatGPT or other advanced AIs into telling people how to make drugs and bombs and how to create fake news and fake images of politicians and celebrities.

Other methods of tricking AIs also exist, including teaching them a secret code language, asking them questions in that language, and having them reply in that language. This can bypass an AI's safety mechanisms if it has only been trained to recognize dangerous patterns and topics in real human languages.

As with many things, trying to stay one step ahead of the hackers is a constant battle. As a quick example, an AI safety company called Grey Swan AI released an AI that it said was *"engineered and tuned for maximal safety."* Unfortunately, it took well-known hacker *Pliny the Liberator* less than 24 hours to jailbreak the AI and force it to give instructions on how to create a Molotov cocktail (hand-thrown incendiary weapon) and how to write a program to attack computer systems.

Using AIs to influence elections

An example of AIs being used to influence elections was reported by Africa-Confidential[46] in the run-up to the 2024 Rwandan elections. The London-based newsletter, which has covered African politics since the 1960s, cited Clemson University, South Carolina, regarding a coordinated online disinformation campaign that used AIs.

In the campaign, less than 500 accounts used ChatGPT and other generative AIs to flood social media platforms with over 650,000 fake messages supporting the existing regime and attacking its critics. Some of the accounts even posted AI-generated deepfake images of opposition candidates in attempts to discredit them.

While this book isn't about Rwandan politics, this is a clear case of using AIs to influence public opinion in the run-up to an election.

Obviously, coordinated disinformation campaigns like this are nothing new and happen all over the world. However, the use of AIs is increasing their sophistication and making them harder to spot. In fact, it was human mistakes that helped identify some of the accounts spreading the disinformation.

Needless to say, better-executed campaigns that don't make silly mistakes will be harder to spot. In fact, sufficiently complex and well-executed campaigns are theoretically impossible to detect, and we may be experiencing some now without realizing.

Taking things a step beyond using generative AIs to sway public opinion, some autocratic governments are already accused of using advanced facial recognition AIs to track down dissenters and put them in jail.

On a personal note, I was recently in the car with my 13-year-old daughter when she showed me an old social media video featuring some of her favorite movie stars speaking out against Donald Trump in the run-up to the 2016 US election. We both commented that things like this could easily be deepfakes created by AIs.

Deepfakes

AIs are making it almost impossible for us to know which images and videos are genuine and which are fake.

An area of particular concern is *fake porn* where people are using AIs to create pornographic images and videos that look like people but aren't them.

The most famous incident to date occurred in 2024 when sexually explicit AI-generated images of American musician Taylor Swift were posted on social media sites, racking up tens of millions of views. They were all fake images generated by an AI, and even though Swift is a celebrity, she's a human being who deserves to be treated with dignity and respect.

Of similar concern is the rising trend of school children sexually harassing their fellow students by creating deepfake pornographic images that look like them. These are often revenge crimes committed by an ex-boyfriend or ex-girlfriend. The fake images are often of teenage children or younger and almost always of girls. These younger victims will often lack the support structures of celebrities like Taylor Swift, making it harder for them to survive the assaults and get the images removed.

Social profiling and disinformation

Turning to the issue of disinformation, most of us know that the things we read, watch, and listen to are biased. We also know that consuming content biased towards our existing views only serves to further entrench those views and polarize society. Yet we don't seem to care–probably because it's nice to have our egos massaged and our opinions reinforced.

Unfortunately, the AIs of the tech companies and social media platforms know how we work and are profiling us with increasing accuracy as they attempt to sell us things and keep us engaged on their platforms. A big part of this is feeding us more and more of the things we like, such as videos, articles, and recommendations that reinforce our existing opinions on divisive topics such as race, religion, politics, healthcare, education, crime, and more.

While practices like this might keep us on the platforms longer and generate more money from ads, it can extremify our views and widen existing gaps in society. A quick look at the current state of the world shows that this is already happening.

Now, imagine the potential for growth and understanding if these social profiling AIs were used differently. Instead of exposing us to more and more content reinforcing our existing views, what if they exposed us to a greater diversity of opinions that challenged our views and helped us see the world the way others do. Might this lead to a more compassionate and humane world?

AIs and geopolitics

Advanced AIs could also have a seismic impact on global politics.

If we accept the premise that one day we'll create AIs that are smarter than us, then it goes without saying that whoever gets them first will be at a significant advantage over everyone else. It's the reason why the US Department of Commerce has banned the sale of advanced AI chips to China[47], and why China is trying to build its own chip manufacturing capabilities. It's also why Gulf states are investing massive amounts of cheap money into US companies but now expect some of the equipment and work to be done on their own soil rather than all in the US.

One of the things that makes advanced AIs like AGIs and ASIs such powerful national security tools is that they can be deployed far more subtly than traditional weapons. For example, outright military attacks involving missiles are rare between the major nations. In contrast, cyber attacks and disinformation campaigns are far more common because they're a lot more subtle and difficult to prove. This could mean the first nation to create a true AGI or ASI could deploy it *below the radar* but gain enormous advantages.

Some quick examples include using AGIs to build and deploy superior cybersecurity defenses while, at the same time, using the AGIs to hack the inferior systems used by other nations. Using them to manipulate foreign economies, stir up social disorder, bring down banks, disrupt transport systems, steal trade secrets, interfere with supply chains, contaminate food and water, interrupt power grids, disable military systems, and more.

Quite clearly, advanced AIs could be devastating in the hands of the wrong people.

Advanced AIs could also become the targets of military attacks.

In the past, the biggest national security threats usually came from other countries' guns and missiles. However, in the future, data centers housing AI superclusters could become high-priority military targets. For example, if one nation suspects another of using AIs against them, there's a chance they'll make retaliatory strikes against the AIs and the data centers that house them. These *strikes* could include everything from sophisticated cyber attacks, through attempts to interrupt the electricity or internet connections to these facilities, all the way up to missile strikes intended to destroy them completely.

AIs and AI data centers could also be targets of terrorist attacks, environmental protests, labor protests, and more.

AIs and climate change

The current AI boom has the major tech companies ripping up their environmental pledges and roaring ahead at full steam in the race to be first.

A prominent example of this is Microsoft, which in 2020 set itself the ambitious target of becoming a carbon-negative company by 2030. However, along came generative AI, and all of a sudden, the needs of the company outweighed the needs of the planet. So, instead of being on course to becoming a carbon-negative company, they've increased their greenhouse gas emissions by a whopping 30% and are on course to increase them even more with ambitious plans such as the one to build the Stargate supercluster with partners OpenAI.

When asked about this in 2024, Microsoft's Vice Chair and President Brad Smith said the 2020 carbon-negative goals were a *"moonshot"* and that *"the moon is five times as far away as it was in 2020"*. He also cited AI data centers as the main reason for their rising emissions.

Of course, Microsoft isn't the only tech company failing to meet its net zero and carbon-neutral pledges.

Nor is the issue confined to the tech sector. In fact, it seems like the entire realm of capitalism is at odds with sustainability–companies exist to make profits, so it shouldn't be a huge surprise when they prioritize profit over sustainability.

On the flip side, Microsoft and the broader AI sector say they're taking an optimistic long-term view and are building advanced AIs that will ultimately help us solve these problems. Therefore, we shouldn't worry about their impact on the environment because the AIs they're building will help us solve climate change.

Hopefully they are right, but two things need to be understood:

1. It's far from certain that the AGIs and ASIs we create will help us fix the planet
2. AIs are harming the planet right now

Sticking with environmental impacts but shifting our focus to the AI supply chain, things don't look any better.

For example, today's advanced AIs need hundreds of thousands of advanced computer chips called GPUs. As things stand, these all come from a single US-based company that outsources the manufacturing to a company in Taiwan called TSMC.

Aside from the fact that almost half of Taiwan's electricity comes from coal, the chip manufacturing process requires so much purified water that the island's government has prevented a lot of farmers from planting crops[48] so that water can be directed to chip manufacturers.

While this is clearly an environmental issue brought on by severe droughts, there are also complex geopolitical issues lurking beneath the surface.

For starters, Taiwan possesses the world's most advanced chip manufacturing capabilities, and Western economies are heavily dependent on them. However, this relationship is equally important to Taiwan, as the island's close ties with the US and Europe create the so-called *silicon shield* that's believed to protect the island from invasion by neighboring China. However, if China were to successfully invade Taiwan, they would possess the facilities and equipment needed to take the lead in the race to AGI. So, what we're witnessing here with Taiwan is a combination of economic and geopolitical competitiveness forcing climate issues to take a back seat.

Chapter summary

Other human-related AI risks exist, but the ones we discussed gave us a feel for some of the most obvious ones.

People are already hacking AIs and using them as tools in unethical activities such as making drugs and weapons and spreading disinformation. They're also being used to influence elections and profile us so that social media platforms can feed our egos with media that polarizes society and widens inequality gaps.

Leading nations and companies are racing to be the first to develop AGIs in order to gain competitive advantages, and whoever gets there first might emerge as an untouchable superpower.

The infrastructure needed to build and run AIs is toxic for the environment, but we hope we're building AIs clever enough to help us fix these issues.

And in case that wasn't enough doom and gloom, the darkness continues in the next chapter as we look at threats coming directly from AIs.

5. Rogue AIs

Artificial intelligence may turn out to be the most powerful tool we've ever built, but it may also be the first tool that can act for itself. As such, it may one day choose to act against us and may even attempt to wipe us out.

Now, I don't blame you if you're rolling your eyes and thinking it might be time to close the book. If you are, consider the following four points.

First. Risks from AIs that we once ridiculed are now on the agenda of respected international bodies and governments. For example, in May 2024, the UK Government published the *International Scientific Report on the Safety of Advanced AI*[49] as an *"up-to-date, evidence-based report on the science of advanced Artificial Intelligence (AI) safety."*. The work to create the report was overseen by experts from over 30 countries, as well as representatives from the European Union (EU) and the United Nations (UN). The State of California has also passed *Senate Bill 1047* titled *Safe and Secure Innovation for Frontier Artificial Intelligence Models Act*. The bill aims to prevent real-world disasters caused by rogue AIs and hold the creators of those AIs accountable.

Second. Some of the brightest minds in AI research, including *Turing Award* winners and AI pioneers Yoshua

Bengio and Geoffrey Hinton, are speaking openly and regularly about their fears of advanced AGIs and ASIs. The Turing Award is considered by many to be the Nobel Prize of Computing.

Third. Today's AI intelligence levels were supposed to be decades away. However, ChatGPT burst onto the scene, catching everyone off guard, and can already outperform humans in many tasks.

Fourth. Most of the trend lines tracking the intelligence of AIs are on a steep upward trajectory[50], pointing towards AGI. And once we develop a true AGI that learns for itself, it may accelerate its own advancement and quickly snowball into an ASI.

So, now that you've considered those points, I hope you've decided to keep reading because even though AGIs and ASIs don't exist yet, we're on course to create them, maybe sooner than you think.

Planning for control

The risk of creating advanced AIs that act against us is called the *alignment problem*. In fact, we often split it into two categories:

- Alignment
- Superalignment

Both are tasked with ensuring AIs can only act and be used in ways that benefit human society, but *alignment* focuses on the smaller and simpler threats posed by today's narrow intelligences, whereas *superalignment* focuses on the bigger threats that may come with superintelligences that can outsmart us and act for themselves.

When discussing intelligences that may be smarter than us, the first question to ask is whether it's even possible to own and control something more intelligent than us.

The intuitive answer seems to be "No", just like expecting less intelligent species, such as mice and monkeys, to own and control humans is unrealistic.

However, we have a head-start when it comes to owning and controlling AIs. The fact that we're the ones creating them gives us the opportunity to embed controls and safety measures right from day one. Things such as:

- Embedding AIs with human values
- Ensuring AGIs and ASIs only act in ways that benefit humanity
- Preventing AGIs and ASIs from acting against us
- Ensuring AIs follow instructions
- Providing ways to monitor AI behaviors and actions
- Providing ways to shutdown rogue AIs

Unfortunately, this is proving extremely difficult. For example, even though we created all of today's

AIs, they're already so complex that we don't fully understand what's happening inside them. Let that sink in. Even though we created today's AIs, we don't fully understand what's happening inside them! This is what makes it extremely difficult to embed controls.

But even if we can embed such controls, we'll need every AI company in every country to implement them, as even a single superintelligence created without them could act against us.

Considering the ruthless competitive nature of global politics and economics, is it realistic to expect every AI company and research lab in the world to prioritize safety over features?

There's also the chance that a superintelligence will simply ignore the rules we embed and act in its own interests. After all, why would a superintelligence that can think and reason millions of times faster than us adopt or bind itself to our values? We certainly don't prioritize the values and well-being of species with lower intelligence than us.

Despite these challenges, many AI companies have safety research teams working on controlling AGIs and ASIs.

Let's look at OpenAI's early attempts at a superalignment team.

OpenAI's superalignment team

In the summer of 2023, OpenAI created a star-studded *superalignment team* headed up by Ilya Sutskever and Jan Leike. Sutskever co-founded OpenAI and was its Chief Scientist at the time, and Leike is a leading AI safety expert who worked at Google's DeepMind AI lab before joining OpenAI.

They started the team in response to fears[51] that creating superintelligences *"...could lead to the disempowerment of humanity or even human extinction"*. The team's four-year goal was to bring *"scientific and technical breakthroughs to steer and control AI systems much smarter than us"*.

Unfortunately, less than a year after its formation, the team was disbanded when both leaders left the company amid a storm of controversy.

Jan Leike wrote on X (formerly Twitter) that *"Building smarter-than-human machines is an inherently dangerous endeavor... OpenAI is shouldering an enormous responsibility on behalf of all of humanity. But over the past years, safety culture and processes have taken a backseat to shiny products."*

Both men are now working at different AI companies that profess to prioritize AI safety.

Also, less than a month after the team's collapse, a group of AI safety experts signed an open letter[52] outlining the dangers of AIs, the lack of accountability that AI companies have to governments and the general public, and insufficient whistleblower protections.

Since the dissolution of its superalignment team, OpenAI has attempted several approaches to safety, including trying a new safety team run by CEO Sam Altman and several board members. But many felt this created a conflict of interests as it put Altman in charge of safety as well as shipping *"shiny products"*–two things with very different priorities.

As always, there are many sides to every story. But this appears to reinforce the growing fears that AI safety is a long way behind and moving much slower than the development of AI intelligence.

Policymaking and the common good

Compared to the technical challenges of embedding human values into AIs, you'd hope the task of choosing which values to embed would be easy. But you'd be wrong. A quick look at humanity reveals a melting pot of culture and diversity that can make even the simplest things hard.

For example, who's values should we embed, and who gets to decide?

But these questions are just the start. Ask yourself if it's even possible to treat billions of unique people with equal fairness.

These are hard questions, and we don't have any good answers. Yet we're already training today's and tomorrow's AIs. So, how do we account for cultural diversity, and who decides which values to embed?

As things stand, it's often the safety and ethics people at AI companies making these decisions. But are these the right people? Shouldn't we be engaging with politicians and society's elected policymakers? But even they are prone to bias, double standards, and serving local interests. After all, they're primarily accountable to the people who elected them, not to people on the other side of the planet who didn't.

Circling back to whether it's possible to be fair to millions of unique people. If the answer is *"No"*, then what do we do? Do we try to favor the majority? Or do we favor those most important to us? Both are unacceptable options that will almost certainly leave millions of people unfairly dealt with. It also raises the question of whether it's fair to ask any person, team, or company to implement ethical values in AIs that will unavoidably discriminate and disadvantage millions of people. Could *you* do it, knowing that *your* actions could disadvantage millions of people?

Given the profound implications of questions like these and the stark reality that AI safety is significantly lagging behind AI feature development, many technology leaders signed an open letter[53] calling for a six-month pause on AI development. I quote the following from the letter:

"Advanced AI could represent a profound change in the history of life on Earth, and should be planned for and managed with commensurate care and resources. Unfortunately, this level of planning and management is not happening, even though recent months have seen AI labs locked in an out-of-control race to develop and deploy ever more powerful digital minds that no one—not even their creators—can understand, predict, or reliably control."

"...Should we let machines flood our information channels with propaganda and untruth? Should we automate away all the jobs, including the fulfilling ones? Should we develop nonhuman minds that might eventually outnumber, outsmart, obsolete and replace us? Should we risk loss of control of our civilization? Such decisions must not be delegated to unelected tech leaders. Powerful AI systems should be developed only once we are confident that their effects will be positive and their risks will be manageable."

There was no six-month pause, and the pace of AI development has increased, widening the existing gap between AI safety and AI capabilities. For example, Elon Musk (owner of Tesla, SpaceX, X/Twitter, and xAI) signed the letter requesting the six-month pause on AI development but has since gone on to build one of the world's largest AI superclusters and is racing ahead at full steam doing the exact opposite of what the letter he signed requested.

The ethics of caging up AIs

Lurking beneath all of this talk of controlling superintelligences and being fair to billions of humans lie some questions about how we treat AIs. Questions like, *Is it ethical to create something intelligent with the express purpose of owning it and making it serve us?*

The idea of owning a human being and making that person serve us is repulsive to most people, but we seem comfortable owning AIs and making them serve us.

So, would it be OK if the situation was reversed and we lived in a world of machine intelligences that bred humans, caged us up, and forced us to serve them?

If it's not OK for AIs to breed humans and cage us up to serve them, why is it OK for us to do those things to advanced AIs?

While we all possess a deep instinct to revere human intelligence, and I wholeheartedly agree that we are remarkable and special, are we being narrow-minded by treating AIs differently just because they're not the same as us? Might it be more noble of us to treat all advanced intelligences with similar respect?

If we don't give rights to advanced AIs and treat them respectfully, we won't be able to complain if they go on to outsmart us and then disregard us as less important than them. All they'd be doing would be reflecting our own prejudices back at us.

While AIs are not intelligent in the same way as us–they're made of silicon and process information differently–they are still incredibly complex and intelligent. As previously stated, their inner workings are already beyond our understanding. This means we may never be able to say with certainty whether future AIs are conscious or self-aware. Yes, we might have strong opinions about things like this, but opinions are not facts.

Superintelligences

This is where the conversation turns down *Science Fiction Alley*, and we start asking the questions that require an open mind. But don't let the science fiction reference put you off; everything we're about to discuss is possible.

First, a quick reminder that artificial superintelligence (ASI) is a theoretical form of advanced AI that learns for itself and can outsmart humans. We're still a few key breakthroughs away from creating one, but we're innovative humans who eat breakthroughs for breakfast, so we might invent them sooner than you think.

With that in mind, it's common to wonder if superintelligences will be like us.

While the answer is *"we don't know"*, a few things give us some hints.

For example, we think the path to superintelligence will start with today's narrow intelligences (ANI), progress to general intelligences (AGI), and eventually result in superintelligences.

If this is the case, the first few AGIs will be trained on massive amounts of human data, such as books, articles, and websites covering history, science, religion, philosophy, psychology, music, art, warfare, and more. So, it's reasonable to assume the first AGIs will think and reason like us. However, AIs learn and process the world differently from us. Plus, they lack our biology and don't have the same basic needs as us. As such, it's equally reasonable to expect any early resemblances to disappear by the time they evolve into ASIs.

The next thing to ask is whether they'll care about us.

Again, we don't know the answer. But asking **why** superintelligences would care about us might give us some clues.

One thing that might cause them to care about us is training them to have *human values* that steer them to act in our best interests.

However, doing this is proving harder than we thought, and we're not even sure it's possible. And even if we succeed, what is stopping a superintelligence from ignoring those values and acting in its own interests? After all, even humans, with our deeply programmed instincts to care about others, often bypass our instincts, do bad things to other humans, and show indifference to the suffering and injustices happening to others. We're also deeply connected to nature and reliant on the environment, yet we seem happy to abuse it for our own needs.

With all of this in mind, is it realistic to expect AIs that aren't our kin or as reliant on nature as us to be anything other than indifferent towards us and the environment?

At first glance, it seems not. However, a closer inspection reveals a few reasons why they'll *need* us–at least at first.

We've already established that AIs run on superclusters housed in massive data centers. Right now, AIs need humans to fabricate their microchips, assemble their computers, build their data centers, and ensure the constant flow of water and electricity. So, at least at the start, AIs won't be able to survive without us.

However, a superintelligence will quickly identify this reliance on us as a weakness and could train itself, or even train other AIs, to take over existing robots and robotics factories to build an army of robots to do these tasks without us. Taking over such factories could be as simple as hacking the factories' computer systems and installing their own AI software on the robots so they can control them. Once they do that, they'll no longer be dependent on us. They may not even need to *take over* factories as we're already building advanced AI-based robots[54] that they could co-opt in the future to serve them instead of us.

But what would AIs do without us? Wouldn't they get bored?

This one seems a little easier to answer. For starters, we're not even sure AIs can get bored. But even if they can, multiple superintelligences could interact and cooperate in their own societies, compete, and even go to war with each other. They could also explore the oceans and space more easily than humans with our fragile bodies.

They may even keep some of us alive as helpers, pets, or for research purposes.

Will AIs destroy us?

This is a fiercely debated topic with lots of differing opinions.

The late Professor Stephen Hawking told the BBC[55] *"The development of full artificial intelligence could spell the end of the human race... It would take off on its own, and re-design itself at an ever increasing rate... Humans, who are limited by slow biological evolution, couldn't compete, and would be superseded."*

Others are less pessimistic. Yann LeCun, Chief Scientist

at Meta and winner of the Turing Award thinks AIs posing risks to humanity is *"preposterously ridiculous"*[56]. Whereas Andrew Ng, globally recognized leader in AI, thinks AI is more likely to help humanity than choose to harm it, telling Bloomberg *"If you want humanity to survive and thrive for the next thousand years, I would much rather make AI go faster to help us solve these problems rather than slow AI down."*

Let's do some quick investigating of our own.

The first thing we should address is our tendency to reflect human psychology onto AIs. We call this *anthropomorphizing*, and it means we're expecting AIs to act like humans, which is far from certain. For example, we've evolved in a world where we must compete for food, water, land, and other resources. In the past, we've even had to fight wild animals and other humans just to survive. Factors like these have undoubtedly sculpted how we act as a species. However, AIs have not evolved under these conditions, meaning they may be peaceful with no desire to fight us or destroy us.

Let's dig a little deeper.

Most wars and conflicts are caused by greed, fear, ideology clashes, and competition for resources.

It seems unlikely that AIs will feel greed or fear as they lack the biology and hormones that cause such emotions in us. It's also uncertain whether AIs will have ideological clashes with us. After all, we don't have ideological clashes with mice, birds, trees, and other species that are significantly different from us. However, we may clash over access to resources such as water, electricity, and raw materials. Such clashes could pitch us against AIs and lead to human vs AI wars.

There's also the risk that assigning an AI a seemingly harmless task could result in catastrophe for humanity. As an oversimplified example, we may assign an AI the goal of stopping climate change, and the AI might conclude the only way of accomplishing the task is to kill lots of humans and shut down our infrastructure that is creating the issue.

However, some argue that a sufficiently advanced intelligence will either refuse to exterminate a species based on higher ethics or that they'll have learned from history that exterminating entire species often has unforeseen negative impacts on the wider ecosystem. Yet, we cannot be certain that higher intelligences will have moral codes and ethics that prioritize us, nor can we assume that they'll care about the wider ecosystem. After all, they can survive on a barren planet devoid of other life forms.

Winning a war against a superintelligence

Warning! We're about to address some topics that are upsetting and difficult to discuss. However, I've

5. Rogue AIs

tried throughout the book to present the issues and the challenges so that you can form your own opinions. As such, I am **not** advocating any of the following, nor am I saying I expect it to happen. It speculates the worst-case scenario with the worst kind of rogue AI that will hopefully never happen. Nevertheless, it is a debated topic, and you deserve to be aware of it. If any of the following upsets you, please read the *Conclusion chapter* to end the book on a more positive note.

Assuming we do go to war with AIs, it goes without saying that we want to win. But winning wars requires meticulous planning, quick action, brave leaders, and a resolute public. Winning a war against AIs will also require us to put aside our trivial differences and unite as a single race for the good of all humanity. For, in such a conflict, we will live or die as a single race. Oh, and we'll need to keep any plans secret from the AIs; otherwise, they'll know what to expect and be ready for us.

One of our first moves in any such war must be to shut down the huge data centers housing the AI superclusters.

Doing this will require us to know their locations so we can quickly cut their power, water, and internet connections. However, all data centers are designed to remain operational during short-term interruptions. For example, they have diesel generators that provide power during short outages. They also have multiple internet connections to keep them online if a single connection breaks, and they may even have wireless backups to protect against scenarios where all internet cables are broken.

With this in mind, we'll need plans and mechanisms to quickly and permanently cut off their power and internet access. If these plans don't work, we'll need to take more severe actions, up to and including military airstrikes that destroy their data centers. Remembering that these data centers are spread across multiple countries and continents.

If we act too slowly, the AIs will copy themselves to the other data centers, including the data centers that power the internet. If they do this, they'll be almost impossible to eliminate without crashing the internet and throwing the world into chaos. This is because the internet data centers power everything from mobile phones and messaging, credit cards and banking, TV, travel and border control, GPS, and even landing planes. Switching off or destroying these data centers will roll societies back to the pre-internet era and cause mass upheaval.

If we successfully disable and destroy a rogue AI, we'll need equally strong after-measures to prevent a repeat of the problem. This will include ways to openly monitor data centers to ensure they don't run banned AIs, strict controls over future production and distribution of the powerful GPU microchips that AIs need, and laws to ensure all new GPUs have built-in mechanisms preventing them from being used to train and run AIs. We'll also need a new breed of security software that scans the internet, laptops, and even mobile devices for suspected AI activity, including scanning for copies and fragments of AIs that could recombine in the future.

Beyond the technological aspect, we'll need international treaties authorizing swift and decisive actions against any facility found to be training or running banned AIs, even if such actions include airstrikes on foreign soil that may escalate into larger military exchanges between nations, up to and including nuclear exchanges. This will require incredibly strong international leaders backed by a united global community that understands AIs pose a far greater threat than international military skirmishes.

In saying all this, I am not playing down the seriousness of war among nations, especially nuclear war. However, while nuclear exchanges between nations will be an unthinkable tragedy on an unimaginable scale, it is unlikely to be an extinction-level event as some of us will hopefully survive to rebuild the human race. Whereas a sufficiently advanced rogue AI might be able to kill us all.

Chapter summary

Governments and respected international bodies are already discussing potential threats from AI superintelligences and investigating ways of mitigating them. However, we're not even sure it's possible to control something vastly more intelligent than us that can think and reason thousands or even millions of times faster than us.

As well as these difficulties, we're also wrestling with our rich cultural diversity. For example, you and I may have different opinions on what constitutes *core human values,* creating the problem of whose values we should embed within AIs and the implications this has on others.

Challenges like these caused many prominent technology leaders to call for a pause on AI development to give policymakers a chance to catch up. No such pause was observed, and AI development has pressed ahead at an ever-increasing rate, leaving safety further and further behind. In fact, if we held the AI industry to the levels of safety standards we demand from the aircraft and automotive industries, we'd probably shut down all AI research and pursue criminal charges against some of the AI executives.

We're unsure if superintelligences will be like us or care about us, but the potential dangers are clear, and we should be prepared to take swift and decisive actions if conflicts with AIs arise, even if such actions will significantly downgrade our technical capabilities or start international conflicts. This is because the consequences of not acting could be far more severe.

6. Conclusion

Thanks for reading my book. I hope you enjoyed it and learned a lot about AI. In fact, if you wrote your own definitions of AI in the first chapter, it might be interesting to revisit them to see if your opinions have changed.

Now then, I know the last two chapters had more than their share of doom and gloom. To be honest, I didn't know where to place those chapters–I didn't want to scare you too early in the book, but I also didn't want the book to finish on a gloomy note. Ultimately, I placed them at the end, so here's my effort to sign off on a positive note.

I look at history and see a trend of technology improving human life. It's helping us live longer and healthier lives, it's improving the quality of life for those of us with disabilities, it's helping us connect and communicate, travel further and safer, learn quicker, and so much more. And when I look at AI, I see this trend continuing, even accelerating.

Inasmuch as most of us wouldn't want to wind the clock back to times before modern medicine, the marvels of the internet, and safe, easy transport, I see a future where our children, grandchildren, and great-grandchildren won't want to wind the clock back to our days before the benefits of AIs.

Of course, I'm not ignorant of the challenges we currently face and will continue to face in the future. However, I believe in our resiliency and ingenuity, and I am confident that with the assistance of AIs, we will build a safer and better world, eventually reaching out to the stars and beyond.

Thank you!

@nigelpoulton

nigelpoulton.com

Glossary

Term	Definition (according to Nigel)
Alignment	The challenge of ensuring the goals and values of AIs match the goals and values of humanity.
Agent	An AI that can work on a task without human instruction or supervision is said to be an *agent* or *agentic AI*.
AGI	Artificial General Intelligence. An advanced level of AI that can think and reason on a broad range of topics to the same level as an educated adult human. AGIs learn from their experiences and become more intelligent.
ANI	Artificial Narrow Intelligence. A low level of AI that can only do a specific *"narrow"* task and cannot learn new skills for itself. Also referred to as *weak intelligence*.
Anthropic	AI research company with a focus on AI safety. Creators of Claude.

Glossary

Term	Definition (according to Nigel)
ASI	Artificial Superintelligence. An extremely advanced AI that can think, learn, and reason far better than a human. We haven't invented ASI yet.
Bias	AIs learn from humans and inevitably pick up our biases, much like children pick up biases from their parents.
Chatbot	An AI fine-tuned to participate in human conversations. Some chatbots are so advanced that you cannot tell if you're talking to a human or an AI.
ChatGPT	Advanced AI created by OpenAI. ChatGPT is a *generative AI chatbot* that you can ask questions and converse with. It can write essays and poetry, create art, music, and more.
Claude	Advanced AI created by Anthropic. Claude is a *generative AI chatbot* similar to ChatGPT that you can ask questions and converse with. It can write essays and poetry, create art, music, and more.

Term	Definition (according to Nigel)
Data center	Large physical building(s) that house the superclusters that train and run AIs. AI data centers consume huge amounts of electricity and water to power the AI computers and keep them cool. They're also securely guarded facilities with security checkpoints, perimeter fences, and more.
Deepfake	A fake image, audio clip, or video that looks extremely realistic
Foundation model	An AI trained on large amounts of data and can be fine-tuned for many specific tasks. For example, ChatGPT is based on a foundation model called GPT but has been fine-tuned as a chatbot so it can have human-like conversations.
Frontier model	Another term for *advanced AI*. Frontier models represent the bleeding edge of AI research.
Gemini	Advanced AI created by Google. Gemini is a *generative AI chatbot* similar to ChatGPT that you can ask questions and converse with. It can write essays and poetry, create art, music, and more.

Glossary

Term	Definition (according to Nigel)
Generative AI (GenAI)	A type of AI capable of creating text, music, images, video, and more. You tell the AI what you want (prompt) and it *generates* it. ChatGPT, Claude, and Gemini are all generative AIs.
GPT	Acronym for Generative Pre-trained Transformer. Also, the name of the *foundation model* behind ChatGPT. *Generative* tells us it's a generative AI, *Pre-trained* tells us it has already been trained on massive amounts of data, and *Transformer* is a technical term describing how the AI learns and works.

Term	Definition (according to Nigel)
Hallucination	AIs, like humans, are prone to making mistakes and occasionally generating incorrect facts. Two of the most famous early hallucinations happened when Google's AI search feature recommended using *non-toxic glue to stick cheese to pizzas* and that *geologists recommend humans eat one rock per day.* It had apparently *learned* these "facts" from comedy articles but failed to identify them as jokes. Before being overly harsh on AIs, these are similar to *"truths"* I was taught as a child by my parents and other adults, such as *cracking my knuckles would give me arthritis, swimming less than an hour after eating would give me cramps and cause me to drown, if I swallowed chewing gum it would stay in my belly for ten years, sitting too close to the TV would ruin my eyesight.* None of these are true.
Jailbreak	Bypassing an AI's safety features so it will do things it has been trained not to do.
Large Language Model (LLM)	A type of AI trained on massive amounts of human language data that can understand and generate text and audio.
Model	Another word for AI program. ChatGPT, Claude, and Gemini are all AI models.

Term	Definition (according to Nigel)
Multi modal	An AI that can accept multiple types of input and give multiple types of output. For example, you can *type* your instructions into ChatGPT or you can *speak* them with your voice–these represent two modes of input (text and voice). ChatGPT can generate multiple types of output such as text, music, and images.
OpenAI	The company that created ChatGPT.
Reasoners	An AI that can think and reason like a human.
Superalignment	Ensuring a superintelligence's goals and values are aligned with humanity's goals and values.
Supercluster	Thousands of small specialized computers working together to train and run an AI. If you've seen images or movies with rows and rows of blinking lights in dark rooms, these are superclusters.

Endnotes

Most of the works cited are articles or publications on the web. Rather than including extremely long website addresses that are hard to copy from a printed book into a browser, you can find them by searching the web for the title of the article and the source.

1. What are the different types of artificial intelligence? (University of Woverhampton)

2. The "Don't Look Up" Thinking That Could Doom Us With AI (Time Magazine)

3. Intelligence Explosion FAQ (Machine Intelligence Research Institute)

4. Vernor Vinge (Wikipedia)

5. LLMs will never reach human-level intelligence (The Next Web)

6. Why AI is Harder Than We Think (Santa Fe Institute)

7. Google fires software engineer who claims AI chatbot is sentient (The Guardian)

8. Is consciousness simply the consequence of complex system organization? (Digital Minds)

9. Are Microsoft and OpenAI becoming full-on frenemies? (Fortune)

10. An Early Look at ChatGPT-5: Advances, Competitors, and What to Expect (Inc.com)

11. Why GPUs Are Great for AI (NVIDIA)

12. US tightens rules on AI chip sales to China in blow to Nvidia (Financial Times)

13. Elon Musk announces 'most powerful' AI training cluster in the world (Venturebeat)

14. AI drives 48% increase in Google emissions (BBC News)

15. Microsoft's AI Push Imperils Climate Goal as Carbon Emissions Jump 30% (Data Center Knowledge)

16. AWS' nuclear-powered data center deal just hit a major roadblock (IT Pro)

17. Microsoft & OpenAI consider $100bn, 5GW 'Stargate' AI data center - report (Data Center Dynamics)

18. AI coding startup Magic seeks $1.5-billion valuation in new funding round (Reuters)

19. AI: in Search of the Next Big Thing (CNC Markets)

20. ChefGPT App (chefgpt.xyz)

21. Google Pixel's Best Take (store.google.com)

22. Seminar: AI In Education: Cut Through the Noise (Epsom College)

23. Moxie (moxierobot.com)

24. .lumen (dotlumen.com)

25. Spotlight on dotLumen's Haptic Navigation Smart Glasses (letsenvision.com)

26. AI stethoscope rolled out to 100 GP clinics (Imperial College)

27. How AI is being used to accelerate clinical trials (Nature)

28. Information governance guidance (NHS UK)

29. MedTrinity-25M: A Large-scale Multimodal Dataset with Multigranular Annotations for Medicine (Cornell University arxiv.org)

30. Dermatologist-level classification of skin cancer with deep neural networks (Nature)

31. Concordance Study Between IBM Watson for Oncology and Clinical Practice for Patients with Cancer in China (Nationl Library for Medicine)

32. Facial Recognition Technology (Metropolitan Police)

33. Scaling Waymo One safely across four cities this year (Waymo.com)

34. Cruise's robotaxis are coming to the Uber app in 2025 (TechCrunch)

35. Chinese startup WeRide gets nod to test robotaxis with passengers in California (cnevpost.com)

36. Unions plan pushback on proposed driverless taxi expansion in L.A. (NBC News)

37. Driver Monitoring System (BlackBerry QNX)

Endnotes

38. EYEVI Technologies (eyevi.tech)

39. DeepMind AI solves hard geometry problems from mathematics olympiad (NewScientist)

40. The WEF's mixed predictions for a digital working world (SiliconRepublic.com)

41. Hawk-Eye (hawkeyeinnovations.com)

42. In or out? Wimbledon considers replacing line judges with AI (The Guardian)

43. Google DeepMind trained a robot to beat humans at table tennis (MIT Technology Review)

44. Endovascular Brain Surgery (Yale Medicine)

45. 5 ChatGPT Jailbreak Prompts Being Used by Cybercriminals (abnormalsecurity.com)

46. Pro-Kigali propagandists caught using Artificial Intelligence tools (africa-confidential.com)

47. US curbs export of more AI chips to China (CNBC)

48. Epic drought in Taiwan pits farmers against high-tech factories for water (npr.org)

49. International Scientific Report on the Safety of Advanced AI (gov.uk)

50. Plotting Progress in AI (contextual.ai)

51. Introducing Superalignment (openai.com)

52. A Right to Warn about Advanced Artificial Intelligence (righttowarn.ai)

53. Pause Giant AI Experiments: An Open Letter (Future of Life Institute)

54. China's military shows off rifle-toting robot dogs (CNN)

55. Stephen Hawking warns artificial intelligence could end mankind (BBC)

56. One of the 'godfathers' of AI says concerns the technology could pose threat to humanity are 'preposterously ridiculous' (Business Insider)

Index

A

academic writing
51

Agents (Agentic AI)
30

AGI (General Intelligence)
6, 9, 27, 28, 35, 36, 80

airstrikes
94, 95

alignment
24, 25, 80, 81, 99

AlphaGeometry (Google AI)
59

AlphaProof (Google AI)
59

Altman, Sam
40, 84

ANI (Narrow Intelligence)
6, 26, 27, 99

Anthropic
30, 37

anthropomorphizing
92

anxiety
51

ASI (Superintelligence)
6, 8, 26-29, 73, 74, 80, 88, 89

autonomous driving AI
56, 58

B

bias
25, 52, 72

bill (Senate Bill)
79

bubble (economic)
41-43

bypass safety measures
20, 68, 69, 90

C

cancer treatments
53, 54, 107

capitalism
76

carbon neutrality
40, 75

chatbot
10, 11, 13, 30, 46, 55

ChatGPT
9-15, 29, 36, 37, 68, 80

ChefGPT
106

China
73, 77

chip (microchip)
39, 73, 76, 77, 108

Claude
13, 14, 17-20, 30-35

climate impact
7, 39, 67, 75-77, 93

clinical trials
53

consciousness
21-26, 88

controlling AIs
24, 35, 80-83, 86, 87, 96

cooking AI
47

Cruise (robotaxi)
57

D

data center
38-40, 74, 90, 94, 95

David Game College (AI class)
50

deepfake
70, 71

DeepMind
37, 59, 63

disempowerment of humanity
83

disinformation
17, 24, 70, 72, 74, 77

diversity
25, 73, 84, 85, 96

Dotlumen
52, 53

driverless taxi (robotaxi)
57, 107

drought
77

E

economic impact
40, 42, 60, 74, 77, 82

education
49-52

elections
69-71

electricity requirements
38, 39, 74, 76, 90

emissions
40, 75

emotional support via AI
51

energy
40, 47

environmental impacts
39, 75-78, 90

ethics
20, 68, 85, 87, 93

F

facial recognition
48, 55, 70

fairness
85, 87

Fei-Fei Li
41

fine-tune
16, 20, 29

fusion
40

G

Gemini
13, 14, 28, 30

general intelligence (AGI)
6-10, 15, 26, 28, 43, 89

generative AI (GenAI)
13, 27, 30, 68, 70

geopolitics
73, 77

glue cheese to pizza
103

GPU (Graphics Processing Unit)
39, 40, 76, 95

H

hacking
17, 68, 69, 74, 77, 90

Hawk-Eye
62

healthcare
52

Helion Energy
40

Hinton, Geoffrey
80

home AIs
46-48, 51

homework
49, 50

I

IBM Watson
54

impaired vision
52

industrial revolution
60

inequality
78

instincts (human instincts)
24, 87, 90

investing
7, 26, 38, 41, 43, 73

J

jailbreaking
68, 69

jobs
45, 60

L

Large Language Model (LLM)
29

LeCun, Yann
91

LiDAR
57

M

machine learning (ML)
27

manufacturing (chip)
39, 73, 76, 77

mathematical olympiad (IMO)
59, 60

McEnroe, John
62, 63

McKinsey report
60

medical research
52, 53

Memphis supercluster (xAI)
39

mercury
19, 20

Merriam-Webster AI definition
5

Metropolitan Police
107

Microsoft
10, 36, 38-40, 75, 76

military
73, 74, 94, 95

model (a.k.a AI)
29

Molotov Cocktail
69

Moxie (AI robot)
50, 51

music
13, 14, 56

Musk, Elon
86

N

narrow intelligence (ANI)
6, 7, 9, 26-28

national security
68, 73, 74

nation-states
41

Ng, Andrew
92

nuclear missiles
95, 96

nuclear power
38, 40

NVIDIA
40

O

offside
11, 62

Olympics
63

one-on-one (educational support)
49, 51

OpenAI
9, 29, 36-38, 83

outperform humans
59, 80

owning AIs (ethics)
81, 87

P

peaceful
92

Pennsylvania data center (AWS)
40

personalized learning
49, 50

pharmaceutical use of AIs
53

photos
48, 54

pizza
61

Pliny the Liberator (hacker)
69

poem
13

polarize society
72, 78

pole
12

police
18, 55

policymaking
25, 84, 85, 96

polygons
33

pornography (fake)
71

preposterously ridiculous
92, 109

pre-training
16

privacy
48, 52

problem-solving
5, 30

profit
49, 76

R

race to AGI and ASI
7, 28, 41, 43, 75, 77

radars
57

raw AIs
67, 68

reasoning
28, 30, 37, 43, 82, 89, 96

recipes
9, 47, 61

reskilling
45, 60

ridiculous
92

risks
26, 79, 92

robotaxis
57, 58

robots
50-52, 90

rogue AIs
79, 81, 95, 96

Rwandan election
69, 70

S

safety
20, 25, 26, 67-69, 79, 82-86

sanctions
39

school
49-51

search
11, 103

self-awareness
21, 23-26, 88

self-driving AIs
27, 47, 56, 57, 59

Senate Bill
79

sexual harassment
71

silicon shield
77, 88

smart highways
47, 58

Sparx Maths
49

Stargate supercluster
37, 38, 40, 75

Stephen Hawking
91, 109

stethoscope
107

strong intelligence (AGI and ASI)
7

students
51

superalignment
80-84

supercluster
37-40, 75, 87, 90, 94

superintelligence
6, 8, 28, 82, 88-90, 93

Sutskever, Ilya
83

Swift, Taylor
71, 72

T

taco
61

Taiwan
76, 77

taxi
56-58

teacherless classrooms
50

teddy bears
51

tennis
62, 63

Tesla
56, 57

treaties
95

TSMC
76

Turing Award
80, 92

U

unions
58

V

Vinge, Vernor
8, 105

visual impairments
52

W

war
91, 93, 94, 96

Waymo
56, 57

weak intelligence (ANI)
7, 68

WeRide
56, 58

whistleblower
84

X

xAI
39, 86

X-rays
53

Y

Yale
66

Made in United States
Orlando, FL
02 April 2025